MYOFASCIAL
TRAINING

MYOFASCIAL TRAINING

Intelligent Movement for Mobility, Performance, and Recovery

Ester Albini

HUMAN KINETICS

Library of Congress Cataloging-in-Publication Data

Names: Albini, Ester, author.
Title: Myofascial training : intelligent movement for mobility,
   performance, and recovery / Ester Albini.
Description: Champaign, IL : Human Kinetics, [2021] | Includes
   bibliographical references.
Identifiers: LCCN 2019052300 (print) | LCCN 2019052301 (ebook) | ISBN
   9781492594703 (paperback) | ISBN 9781492594710 (epub) | ISBN
   9781492594727 (pdf)
Subjects: LCSH: Fasciae (Anatomy) | Kinesiology.
Classification: LCC QM563 .A43 2021 (print) | LCC QM563 (ebook) | DDC
   611/.74--dc23
LC record available at https://lccn.loc.gov/2019052300
LC ebook record available at https://lccn.loc.gov/2019052301

ISBN: 978-1-4925-9470-3 (print)

This book is a revised edition of *Allenamento Mio-Fasciale,* published in 2018 by Elika Editrice.

Editing: Marta Viola Rustignoli and Julie Marx Goodreau; **Layout:** Claudia Peroni and Julie L. Denzer; **Cover design:** Sarah Bocconi and Keri Evans

Images: The images on pp. 14(left), 19(bottom), 24, 25, 26, 28, 29, 30, 35, 38, 42(top), 44, 47, 48, 53(top), 54, 55, 57 and 228 are from Novagram, Luca Leurini, Graphic Designer, Rimini. p. 13: Bearacreative/Shutterstock; pp. 14(right), 20: Liu Zishan/Shutterstock; p. 15: Designua/Shutterstock; p: 16(left), Teguh Mujiono/Shutterstock; p. 16(right), ively/Shutterstock; p. 23: Blue Ring Media/Shutterstock; p. 31: Alila Medical Media/Shutterstock; p. 32: Miha de/Shutterstock; p. 33: Designua/Shutterstock; pp. 39, 68: Lotan/Shutterstock; pp. 50(top), 66: Sebastian Kaulitzki/Shutterstock; p. 50(bottom): udaix/Shutterstock; p. 56: Hidemozart/Shutterstock; p. 65(woman): Ersler Dmitry/Shutterstock; p. 65(kite): Cindy Lee/Shutterstock; p. 65(bottom): nogandosan/Shutterstock; p. 69: EniaB/Shutterstock; p. 108: Scio21/Shutterstock; p. 132: Alexonline/Shutterstock; p. 152: Elena Butinova/Shutterstock; p. 209: Aspen Photo/Shutterstock; p. 229: Sanit Ratsameephot/Shutterstock; p. 275: ellepigrafica/Shutterstock; p. 277: Kjpargeter/Shutterstock

Printed in the United States of America   10 9 8 7 6 5 4 3 2 1

The paper in this book is certified under a sustainable forestry program.

Human Kinetics
1607 North Market Street
Champaign, IL 61820
Website: www.HumanKinetics.com

In the United States, email info@hkusa.com or call 800-747-4457.
In Canada, email info@hkcanada.com.
In the United Kingdom/Europe, email hk@hkeurope.com.

For information about Human Kinetics' coverage in other areas of the world, please visit our website: **www.HumanKinetics.com**

Tell us what you think!
Human Kinetics would love to hear what we can do to improve the customer experience. Use this QR code to take our brief survey.

E7992

CONTENTS

ACKNOWLEDGMENTS

Each of us has a pair of wings, but only those who dream learn to fly.
Jim Morrison

I soar through the skies with the wings that I dream of having, to be a little closer to you, Dad.

I soar through the skies with the strength and power of my wings for you too, Gigi.

I continue to soar through the skies for everyone who has believed and continues to believe in me.

Thank you so much... I shall never tire of flying.

For Jacob and Nicole: I wish nothing less for you than to spread your wings and fly freely.

For Guido, my pilot and flight companion who is by my side and always guides me back down the right path: I love flying with you to magical, unknown places and through the storms. The only thing that matters is that you are by my side. May our journey never end.

Thanks to my faithful students, who have been following me for years: Ugo, Loredana, Isabella, Simona, Novella, Benedetta, Paola, Anna, Susi, Cristina, Iva, Silvia, Luana, Monica, Francesca, Valeria, Antonella, Giovanna, Ilse, Sara, and Gigi. Many thanks to you too, because you are part of my journey.

Thanks to the fans of my training courses: to Nora, Isabella, Antonietta, Mimi, Laura, Carmen, Anna, Gisela, Sabina, Giovanna, Elena, Antonella, Michela, Agnese, Paolo, Daniela, and everyone who is following me.

Thanks to my great friend Jacky Klossner, who is far away and yet always close.

To Karin Locher, who opened my eyes to a fascinating world about myself: Thank you from the bottom of my heart.

Finally, thanks to Elika, who had the patience to wait for my book.

INTRODUCTION

FReE: Fascial Real Emotion

FReE (Fascial Real Emotion) is much more than just movement. It is the key that will enable you to change the "fascial" clothing that you wear every day: You will instantly feel younger, more elastic, and more energetic. Not yesterday, not tomorrow, but now is the right time to change. What are you waiting for?

In this book I will offer a step-by-step guide, starting with simple movements. In order to change, serious commitment shall be required on your part to put the exercises into practice and follow my recommendations; otherwise you won't see the desired results.

The exercises proposed are easy to do and have a significant impact on the fascia, but it can be a challenge to repeat them day after day. If you have a goal, don't lose sight of it and the results will come. Shrug off your old routine and you will reach new heights!

The New Key to Intelligent Movement

Throughout my 30 years of experience, my goal has always been to make people feel good, with their body and mind together, inseparable, as they really are. That is why I have always looked to improve my understanding of all the various aspects of human movement.

I was a Les Mills, Reebok, Polar, and Airex BeBalanced trainer for many years, as well as a fitness instructor and personal trainer. I have attained qualifications in posturology (Bricot, Mézières), Pilates (Polestar, Balanced Body, Fletcher, Hermann, Locher), the functional method, kettlebell training (London, IKFF), and Gyrotonic, going to the Unites States every other year to consolidate my skills. Each of these courses added new strings to my professional bow and to my training. However, I was still me, just with a few more materials, methods, and professional contacts to call upon—and a little more experience.

I first discovered the myofascia thanks to a book by Thomas Myers. To be honest, I didn't understand much of it, but I liked it. Things never happen by chance, but what matters, happens. This really is true! I followed my intuition.

I started moving from place to place, taking part in courses by Thomas Myers (USA, Switzerland, Germany, the Netherlands), by Michol Dalcourt (USA and the Netherlands), by Robert Schleip (Germany), and by Karin Locher (England and Switzerland). Each discipline does of course take time to be applied, absorbed, felt, and perceived. I entered the myofascial world on tiptoes, enriching it with classic Italian cheerfulness and light-heartedness—and above all living it.

And that is where the magic happened. I don't know how else to explain it. In a random place, on an afternoon like any other, I was in Zurich, and I felt something change deep within me for the first time; it was as if a block that I hadn't been aware of had been removed from me. Can I admit that? I went from euphoria to amazement and then to fear... and I cried.

That afternoon I realized that something had happened; I felt myself change for the first time, and my own perception of myself was no longer the same. In a word, I was free.

That is where the name FReE (Fascial Real Emotion) comes from, which explains in a single acronym what myofascial training really is all about. By working on the body more deeply than we are used to, we can understand things that are normally ignored and go unnoticed. By feeling my body and experimenting with a new way of using it, I discovered the meaning behind what I was doing and realized I had found a new key to movement.

Welcome to the world of FReE. Happy reading!

Stay connected and FReE!

Ester Albini

WHAT IS FASCIA?

1.1. FASCIA AND THE FASCIAL SYSTEM

1.1.1. Background

A spider's web in the morning dripping with dew. This is how I suggest you think of the fascia.

It is a biological web. If you could remove your outermost layer of skin, you would see the fascia as a close-fitting, white, extremely thin and light web that covers the whole body from head to toe.

When I ask people to think about their own bodies and the first anatomical image that pops into their heads, they talk about bones, muscles, and internal organs. No one, or practically no one, visualizes an almost transparent system that is the fascia, a system that in recent years has attracted the attention of movement therapists, osteopaths, and manual therapists.

Why is this image lacking?

One answer lies in the origin of the word "anatomy", which derives from the Greek ανατομή, meaning "dissection" (ανά – through, and τέμνω – cutting or separating the parts). This etymology explains how we came to produce such detailed anatomy books and to acquire today's beliefs about the structure of the body: by dissecting dead bodies.

Let's take a step back in time. Dissecting cadavers was the only way known to the first anatomists to conduct exploratory research on the human body. In reality, the work of anatomists, including Italy's Leonardo da Vinci, was an important foundation for the study of the human body, and these discoveries would influence our beliefs for centuries. At the time, the tools required for this type of study had not yet been invented, and it was only much later that scalpels and other precision instruments became available.

Dissections were therefore performed to better understand the human body. Unfortunately, these pioneers cut everything in between one tissue and another without dwelling on the fascial tissue, unwittingly overlooking its vital role. As a result, the very concept of the human body has been intellectually linked for centuries to this division, which considers the various systems to be separate entities, and they appear that way in every anatomy textbook. For centuries, we have focused on examining the individual parts without understanding that they only function in constant and mutual cooperation as a unified system. My opinion is that, as in other parts of life, we tend to look at the bigger things before concentrating on the smaller things. But although they are small, they are no less important.

1.1.2. Defining Fascia

Fascia is what makes us who we are. Connective tissue is simply everything that binds and connects.

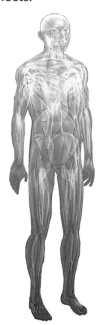

However, experts cannot agree upon a single definition. There are two key reasons for this lack of consensus on the fascia:

1) Its thickness, function, depth, consistency, and position are extremely variable.
2) It is a concept that has only been broadly accepted relatively recently, so its definition is still a work in progress.

The simplest definition that I have found is this: "Fascia is a bandage", a concept that already existed in documents from the 18th century [15].

The fascial tissue is a four-dimensional network that wraps around and separates every part of the body, creating a structural continuity that gives shape and function to all tissue and organs. The human body is a functional unit in which every region is in communication with another through the fascial network.

> Fascia can be defined as a four-dimensional structure because it extends beyond the three physical dimensions (width, height, and depth) to include a neural or sensory dimension, which represents its close connection with the central nervous system.

Think of the fascia as a close-fitting and semitransparent web that wraps us and connects us from head to toe and acts as an external nervous system that processes and responds to sensory and mechanical stimuli.

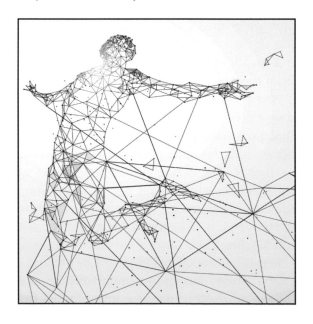

The fascial tissue, which can be found throughout the body, surrounds and permeates blood vessels, nerves, organs, the meninges, bone, and muscles; interacts with them; creates various layers at different depths; and forms a four-dimensional matrix of mechanical, metabolic, elastic, and neurovegetative characteristics. It is four-dimensional because it operates beyond the three physical dimensions to connect and interact with the central nervous system: it's a genuine fourth dimension. In this light, the fascia becomes an organ that affects a person's health. From a more general standpoint, knowledge of its functions and of the areas it controls becomes important for a person's health and well-being.

If we could observe the fascial structure separately from the rest of the body, we would see an extremely dense four-dimensional web, with no beginning or end, that separates, connects, and gives shape to everything. It is a semitransparent network that starts from the skin and thickens in the fibrous tissue that surrounds muscles, bones, and the internal organs. It is a continuous system that covers and crosses our body and accounts for 20 percent of our body weight.

To simplify the concept, I would like you to consider an orange. I use this as a visual aid in my training courses, comparing it to our fascial system. Take an orange and cut it in half. If you look at the cross section, you will see the individual segments separated by white skin.

Just as the orange is surrounded by a white skin of cellular tissue, which simultaneously maintains the consistency and gives shape to the pulp, our body (under the layer of skin) is surrounded and covered by connective tissue called superficial fascia. In addition, the orange is divided into segments that contain small sacks full of juice. The human body is very similar, since every structure of the body, every muscle, and every organ is surrounded by a sheath of connective tissue. Even the

juice of the orange can be compared to the ground substance (aqueous gel) found in the body (see section 1.3.1).

Let's take a journey into our body, from the surface (skin) to deep within (bones), passing through the different layers. The first superficial layer of subcutaneous fat, intertwined with the first layer of connective tissue, the so-called **superficial fascia**, can be found under the skin (dermis). The **deep fascia** is located after the deep layer of fat. Continuing our journey, we reach the **epimysium**, a layer that wraps around the whole muscle; the perimysium, which covers the muscle bundles; and the endomysium, which covers every fiber and muscle cell. Finally, we come to the periosteum, the layer that covers the bone.

The image below shows the different fascial layers.

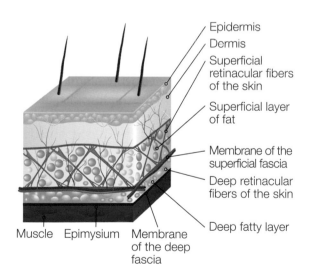

Returning to the analogy of the orange, the white skin represents the fascial components. Removing the white skin would just leave the juice of the orange. In the human body, the principle is the same: If the fascial layers (epimysium, perimysium, and endomysium) were removed, the muscle would lose its shape and consistency.

Visual Comparison of an Orange and a Muscle

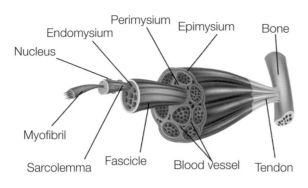

Cross section of a muscle

Cross section of an orange

The epimysium, perimysium, and endomysium can be clearly distinguished in the figure on the left.

- **Epimysium**: Wraps around the whole muscle.
- **Perimysium**: Covers the bundles of muscle fibers and connects them to form the most abundant fascial tissue of the body. Conducts blood vessels and nerves to muscle bundles (control of the perimysium's nutritional function). It is a mobile layer that, during contraction, allows the muscle to slip inside its casing.
- **Endomysium**: Surrounds every muscle cell, creating an individual unit. It is arranged in tubes that wrap around every muscle fiber.

Everything I learned during completion of the Fascial Dissection certification with Thomas Myers completely changed my perspective of human anatomy, given that until then I had only studied it in traditional anatomy textbooks. These entail nothing more than colored illustrations of muscles in which the fascia is not even visible and is therefore not given the importance it deserves. The individual muscles are depicted with an insertion and an origin and are represented as separate from one another, but this does not reflect reality. In fact, every muscle is connected to its adjacent muscles and communicates with them through the fascia.

Try this with the orange: If you want to separate the segments from each other; you will have to peel off the white skin. The segments are not just tightly packed against each other; they are attached and connected to each other. The same is true of our muscles.

This therefore represented a hugely significant discovery, since it affects how I now see the body as well as my approach to training. Remember: The fascia separates, gives shape, and communicates.

It should also not be forgotten that the white skin of the orange is made of fluids and fibers. Taken as a whole, this structural model ensures high tolerance to deformation compared to an apple, for example. Pressing an apple with your finger creates a permanent pressure point. However, if you exert light pressure on an orange, it deforms but will return to its original shape over time.

The fascial tissue performs the following functions.

Connecting Function
The fascia acts as a "placeholder" for the muscles and organs and therefore stabilizes the body.

The entire bone system is in contact with connective tissue, joint capsules, and ligaments. The muscles are connected to bones (periosteum) by their tendons. The muscles, organs, and skin are connected to the surrounding tissue by fascial structures. It is a four-dimensional network that encloses the whole body and has no beginning or end.

Perception of the Body

The body is able to perceive thanks to sensory communication, which relies more on the fascial structures than on the joint and muscle structures. The receptors responsible for our perception of the body are up to six times more abundant in the fascia than in the muscles. This is extremely important for accelerating the healing process, increasing well-being, and enhancing performance.

Impact on Flexibility

The fascial tissue network connects everything with everything else. If the fascia is well-hydrated and elastic, its adjacent structures can slide freely over each other. Dehydrated fascial tissue, on the other hand, has a negative impact on flexibility, thereby reducing well-being and sporting performance and increasing the risk of injury.

Quality of Movement

The fascia is involved in every movement we make. The quality of a movement depends on the structure of the muscles and of the fascia, as well as their coordination.

Transmission of Kinetic Muscle Energy

It has been discovered that the energy generated by the muscles is transferred to other parts of the body during a movement, not only by the ligaments, tendons, and joint capsules but also and predominantly thanks to the fascial structures that surround the muscles. If the transmission properties of the fascia are good, an athlete can achieve his or her maximum performance. However, if the fascia is not trained, these properties may be altered and inhibited, with an increased risk of diminished performance and injury.

Defense

Fascia plays a defensive role from an immunological perspective. In fact, our body's immune system relies on the quality of the fascia.

In a well-balanced and healthy fascial structure, waste is transported away. Many phagocytes can be found in the ground substance, which is similar to a transparent gel and surrounds all the cells in the body. These phagocytes act as garbage collectors that ingest and destroy cellular debris and bacteria. In dehydrated fascia, a lack of fluid due to a lack of movement or unilateral movement inhibits the function of many of these cells that have specific functions, which literally remain dry.

Transport and Nutritional Function

Nutrients are transported from the arterial system via the connective tissue to where they are needed; in the other direction, waste is carried via the connective tissue to the venous vascular system or the lymphatic system.

Cause of Many Types of Pain

The fascial structures contain many pain receptors. Many scientists now subscribe to the theory that about two-thirds of all pain is associated with the fascia. Numerous studies have shown a direct link between myofascial pain and the body's perception. For persistent myofascial pain, the body's perception in the painful region is significantly reduced. However, if the body's perception in this region is improved, the myofascial pain diminishes or completely disappears. These studies found that many people reported reduced pain after a training session with a foam roller, balls, sticks, and other similar equipment.

1.1.3. Fascia Is Connective Tissue

Anatomically speaking, the term "fascia" means a membrane of fibrous connective tissue made of collagen fibers and elastic fibers that seamlessly covers all the musculoskeletal structures of the body and biomechanically connects them to each other. It is a network of fibrous protective tissue for the body's organs, systems, and complexes that interacts with the body at every level, from the surface to deep within. In reality, the fascia's connection with all the body's structures is such that defining it simply as connective tissue does not do it justice.

The fascial system is an organ or system that provides muscular-intramuscular, organ, and inter-organ support and connections. The fascia surrounds every structure of the body and links and communicates with all the endogenous systems. These functions are related to the synchronization of movements between muscles, organs, blood vessels, and nerves.

Scientific research has shown that the fascial continuum is innervated by the autonomic sympathetic nervous system. The fascial continuum plays a crucial role in transmitting muscle strength, ensuring correct motor coordination, and holding the organs in place. The fascia is an essential element that allows us to communicate and live autonomously.

The various structures that make up the human body often have to respond to mechanical stress or perform moving or sliding functions. The connective tissue, which is responsible for supporting these movements, must have a structure that is sufficiently elastic to respond to these stresses and, once resolved, to return the structure to its original size and shape.

The term "connective tissue" is used to describe "a set of different tissues that share a common origin from the embryonic mesenchyme. The mesenchyme is embryonic tissue from which all connective tissue originates. It is composed of cells that have no specific characterization and can differentiate into the different cell types that make up the connective tissue" [24].

There are three types of connective tissue:

1) **Connective tissue proper**: This includes loose connective tissue, dense connective tissue, reticular connective tissue, and elastic connective tissue.
2) **Specialized connective tissue**: This includes adipose tissue, cartilage tissue, bone tissue, and blood.
3) **Embryonic connective tissue**: This includes the mesenchyme and the mucosa.

In this book we shall only focus on the connective tissue proper, which is further divided into two sub-classes according to the density, proportion, and orientation of its fibers, and the types of cell it comprises:

1) **Loose connective tissue**
2) **Dense connective tissue**

"In the first, the fibers are loosely intertwined, while in the second they are very abundant and gathered in large clustered bundles that give the tissue a remarkable consistency" [44].

Connective tissue "fills all the free spaces between the organs, inserting itself between them and connecting them together" [45]. Loose connective tissue is primarily made up of ground substance (an aqueous gel). This incorporates collagen fibers and thin bundles of elastic fibers, which are less abundant and loosely woven together [14]. In terms of its function, it is not just used as a filler but also as a water reservoir, a sliding layer, and as a living space for numerous free cells.

Dense connective tissue is characterized by a large number of collagen fibers. As a result, it contains less ground substance than loose connective tissue. It can be further subdivided in accordance with the structural direction of the collagen:

1) **Reticular dense connective tissue**
2) **Parallel dense connective tissue**

The fibers may be arranged in a reticular pattern (figure A) – wavy collagen fibers and elastic fibers, arranged in a transverse, longitudinal, and oblique direction – such as in the dermis. Or they may be grouped together in parallel bundles (figure B), such as in the tendons, ligaments, and aponeuroses (regular dense connective tissue).

Finally, some types of loose connective tissue have special properties: mucosal tissue, elastic tissue, reticular tissue, adipose tissue, and pigmented tissue.

Reticular dense connective tissue (figure A) is largely elastic and possesses the following characteristics:

- Arrangement of the fibers: Reticular; this enables it to stretch far in all directions.
- Site: Visceral organ capsules, dermis, periosteum and perichondrium, nerve and muscle sheaths; however, this reticular structure can also be found at flat parts such as the plantar fascia (on the sole of the foot), the lumbar fascia, and the muscle sheaths.
- Function: Exerts resistance to oppose forces applied in multiple directions; prevents hyperexpansion of the organs.

In contrast, parallel dense connective tissue (figure B) is made up of largely non-elastic fibers and is characterized as follows:

- Arrangement of the fibers: Parallel; this gives rise to a stabilizing action.
- Site: Between the skeletal muscles and the skeleton (tendons and aponeuroses), between the bones (ligaments), at the joint capsules, around the skeletal muscles (deep fascia).
- Function: Provides a robust link between tendons and bones; coordinates muscle traction; reduces friction between muscles.

The figure below illustrates the different classifications of connective tissue. Considered here to be fascial tissue, they differ in terms of density and directional alignment of the collagen fibers [60]. The closer the fascia is to the surface, the less structured its organization is.

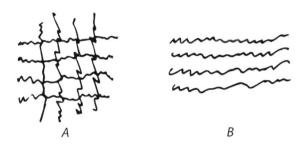

A   B

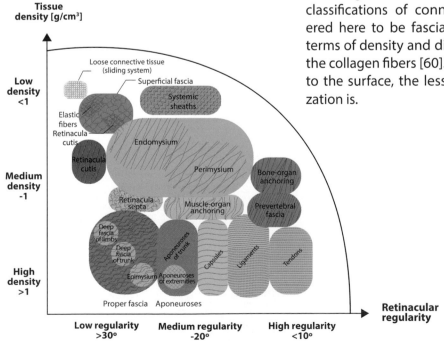

Classification of body tissue by density and regularity of the connective tissue.

1.1.4. Anatomy of Fascia

The anatomy of fascia is a new branch of anatomy itself.

Even though we often refer to *one* fascial network, the fascial structures are in fact differentiated into four key areas according to their position and function. Although there is little consensus regarding these classifications, in this book I have drawn inspiration from the model proposed by F. Willard [76]: The fascia can be considered to be subdivided into four layers forming interconnected concentric longitudinal cylinders.

1) Superficial fascia
2) Deep fascia
3) Visceral fascia
4) Meningeal fascia

Superficial Fascia

Think of the fascial layer as a wetsuit that covers the whole body, keeps it together, supports it, and gives it shape.

The superficial fascia is the outermost layer covering the whole body; it is located directly below the skin and is intertwined with subcutaneous fat. It consists of different levels, each with varying numbers of fibroblasts

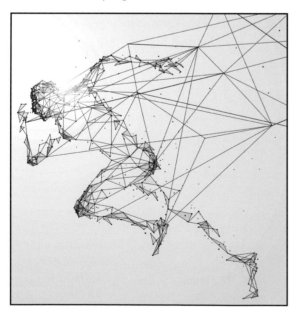

(connective cells) arranged in a reticular pattern (multidirectional).

The superficial fascia is full of water in the form of liquid crystals and is composed of loose connective tissue (subcutaneous, which may contain woven collagen fibers and larger quantities of elastic fibers) and adipose tissue.

This fascia is an important storage location of water and fat, it protects against mechanical and thermal deformation (insulating layer), it is a path for multiple nerves and blood vessels, and it allows the skin to slide over the deep fascia. The latter is particularly apparent at the highly mobile joints and on the back of the hand, where the skin is able to move freely and easily over the extensor tendons when you move your fingers. Mobility is facilitated by the presence of multiple layers of collagen fibers coupled with elastin.

The predominantly reticular (multi directional) arrangement of the fibers enables the fascia to stretch in all directions. The superficial fascia is connected to the deep fascia by loose connective tissue.

Functions:

- It determines the body's external appearance.
- It is very adaptable and provides an ideal surface for sliding.
- It can be stretched in all directions.
- After stretching, it returns to its starting position.

Deep Fascia

The deep fascia is the last connective layer before coming into contact with the body itself (i.e., bones and muscles) and with the organ and vascular systems. It consists of overlapping layers of reticular dense connective tissue (wavy collagen fibers and elastic fibers arranged in a transverse, longitudinal, and oblique direction) that exhibit various biomechanical characteristics at different levels. It is a rather cohesive cylindrical layer around the body (trunk and limbs), and it also forms a membrane that covers the outer part of the muscles.

A distinctive feature of the deep fascia is that it comprises structural and functional compartments that contain particular muscle groups with specific innervation. The compartment also confers the muscle's specific morphofunctional characteristics: A muscle that contracts inside a sheath develops a pressure that sustains the contraction itself.

In an individual muscle, through the septa, aponeuroses, and tendons, the deep fascia continues to the **muscle fascia**, which comprises the epimysium (fibroelastic connective tissue that wraps around the whole muscle) and extends into the muscle belly. The muscle belly itself is formed of the perimysium (loose connective tissue that covers the fascicles of muscle fibers) and the endomysium (delicate connective sheath of the muscle fiber). In normal conditions, these septa and sheaths both nourish the muscle fibers and allow them to slide.

This fascia is directly linked, both anatomically and functionally, to the neuromuscular spindles (muscle spindles) and to the Golgi tendon organs. It also possesses well-characterized receptor properties (see section 1.8).

Functions:

- It wraps around the muscles and connects them to each other, organizes them into functional units, and allows them to slide over one another. It is very adaptable.
- It is composed of tendons, ligaments, and other structures; it provides sensors with feedback on its position and movement, as well as the movement of individual muscles and neighboring structures.

Visceral Fascia and Meningeal Fascia
Every organ is surrounded by a fascial layer that, thanks to the ligaments of which it is formed, maintains the organ's structural position and its integration in the body's fascial network as a whole. Think of the fascia (white layer) that covers the meninges, the "bag" that surrounds the heart, and the film that covers the parietal pleura of the lungs (endothoracic fascia).

Functions:

- It delimits the adjacent structures and allows the organ to slide.
- It connects an organ to its surroundings.

From both an anatomical and functional perspective, connective fascia and muscles constitute the myofascial system, playing a key role in balance and posture. Fascial tissue is in fact the body's largest sensory system. It sends signals to the central nervous system and boasts a large number of mechanoreceptors up to the visceral ligaments and spinal and cerebral structures.

Mechanoreceptors able to induce local and general effects have been found in abundance in the fascia up to the visceral ligaments and in the dura mater surrounding the brain and spinal cord (dural sac) [11].

In addition, the fascia contains an abundance of proprioceptors (notably Pacini and Ruffini corpuscles), particularly in the spaces between joints and fascia and between fascia and muscle. The fascial continuum is a bona fide sensory organ that plays a role in everyday movement and posture.

To summarize the characteristics and functions of the fascial system, the fascia surrounds and separates, supports and lubricates, and adapts itself plastically and elastically to the forces that pass through it. We owe our shape to the fascia – it is what makes us who we are.

1.2. TENSEGRITY

Tensegrity is the ability of a system to maintain mechanical stability with the tension and compression forces at play, which are distributed and balanced. The slightest variation in localized tension is transmitted around the body by the fascial system. Literally, the word "tensegrity" is a portmanteau of **tensional integrity**.

ANATOMICAL TERMINOLOGY

Tendons and Ligaments

If you can imagine a string of fresh sausages, the connections between them are the tendons. This image helps us to understand that the tendons are not individual links between the bones but rather a part of the fascia surrounding the muscles or organs that performs a different function.

Ligaments are fascial structures that connect bones together.

Aponeuroses

The term "aponeurosis" refers to a thin, fibrous fascia that covers and wraps around the muscle. Aponeuroses are arranged in sheets of various thicknesses and tend to have a large surface area. In practice, they act as tendons that are sheet-shaped, while the actual tendons are shaped more like a band. On one side the aponeuroses continue directly into the muscles, while on the other side they insert into skeletal segments and join together to form a tendon (like in the Achilles tendon).

Aponeuroses are arranged in different directions to the adjacent planes to guarantee resistance to traction. Aponeuroses are thickest and most evident in the ventral abdominal and thoracic-lumbar regions, as well as the palmar and plantar regions.

From a mechanical and myofascial perspective, aponeuroses are the mechanical agent of motor coordination wherever strong and explosive force is required, like a catapult (for more information about the catapult effect, see section 7.3).

Septa

These are fascial membranes that separate two muscles or two groups of muscles. For example, in the thigh, an external intermuscular septum and an internal septum separate the anterior compartment of the extensors from the posterior compartment of the flexors.

Cartilage and Joint Capsules

Cartilage tissue is a special supportive connective tissue. It is composed of cells called chondroblasts and chondrocytes, immersed in and synthesized by abundant extracellular matrices formed of collagen fibers and a gelatinous matrix.

The joint capsule is a fibrous and resistant structure that protects and maintains *in situ* the entire joint. The bone ends (articular surfaces) are held close to one another but are not joined together. The joint capsule is subdivided into two layers:

1) Outer fibrous layer, directly connected to the periosteum
2) Inner layer (the synovial membrane), which helps to delimit the joint cavity, filling it with synovial fluid

Architectural models have been constructed that recreate this concept: flexible elements that react to traction and rigid elements that react to compression forces, all so that the structure can self-stabilize. An example of this is the geodesic dome designed by the American engineer and architect Buckminster Fuller [8]. Domes adapt easily to predictable and unpredictable forces to varying degrees depending on the materials used [8]. The three-dimensional model is composed of more tie rods and rigid elements and can become deformed and return to its original neutral position irrespective of any force it may be subjected to from any spatial direction, whether compression or traction.

This model seems to be a good representation of the cells in our body and, more evidently, of the connective tissue and the myofascial system in particular. The myofascial system affects body movement and stability, because it works through different forces acting on it, such as traction, flexibility, compression, and torsion. All types of movement and motor input become a message or a request (whether stretching, support, breathing, or contraction) to which the fascia responds. This mechanism furnishes the myofascial system with a new awareness, creating new, more advanced conditions to enable our body to adapt to a new request.

Let's take a dome tent as an example. After erection, the tent holds its shape. This is known as preload or pretension, which refers to a balanced distribution of forces in all its parts through every individual

structural component. The pretension of each structural component as an individual unit is carefully calibrated during the manufacturing process in order to resist the external and internal forces once the tent has been put up. *Tensegrity*, an amalgamation of tensional integrity, is established from the combination of compression forces with those that adapt to traction forces. Tensegrity or balanced preload enables a structure to absorb gravitational forces, vibrations, wind, and so on, without sustaining damage. That is why the best tents for extreme atmospheric conditions, such as those used by trekkers, are all dome tents of one form or another. The human body also uses this strategy, and there is clear evidence to suggest that all our cells are biotensegrity structures.

1.3. STRUCTURES THAT MAKE UP THE FASCIA

1.3.1. Structures That Make Up the Fascial Tissue

Connective tissue is composed of many different structures, the function of which will be explained in this chapter. The aim is to better understand the changes that arise from training the fascia.

Fascia is made up of fibers and cells that are immersed in (and crossed by) ground substance, which is like a transparent gel. The reticular and elastic collagen fibers are found in the ground substance, which is also known as the extracellular matrix (intercellular fluid). Apart from water, the other primary component is collagen. The latter are bulky molecules comprising a main protein axis, known as a *core protein*, into which complex polysaccharide chains, called glycosaminoglycans (GAG), are inserted.

Structures That Make Up the Fascial Tissue

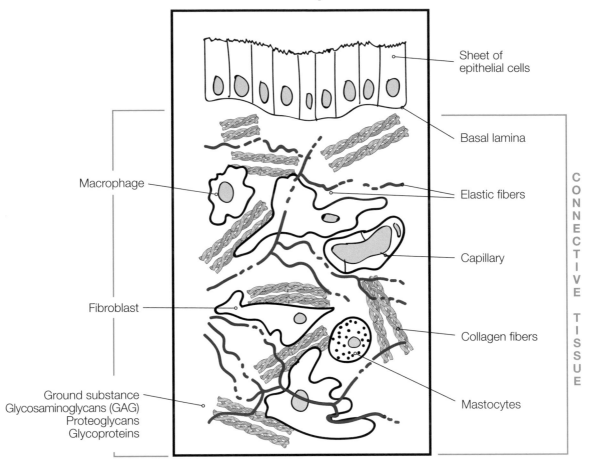

The main components of fascial tissue are

1) ground substance (also known as the extracellular matrix),
2) fibers,
3) cells, and
4) water and nerve fibers (water is discussed in this section, while nerve fibers are addressed in section 1.6).

Ground Substance

The cells and fibers of fascial tissue are immersed in a viscous fluid similar to very watery gelatin called "ground substance". This liquid, which fills all the space between the cells and the fibers of the connective tissue, is composed of long and intertwined proteoglycan molecules, which themselves are formed of chains of glycosaminoglycans (GAG). These are able to bind to water, thereby making the matrix permeable to metabolic substances and gases that pass from the blood to the cells of the tissue and vice versa.

As such, the ground substance is where significant exchange activity takes place, while it also acts as a shock absorber, resists compression, lubricates, and regulates intercellular exchange. The amount of water it contains accounts for around 70 percent of the entire connective tissue.

Fibers

Fascial tissue consists of three basic types of fiber:

1) Collagen fibers
2) Elastic fibers
3) Reticular fibers

The structures of the individual healthy fibers are defined as *crimped*. The fact that they are very wavy enables the tissue to stretch if required before returning to its starting position. As you will remember, the structural arrangement of fibers may be reticular or parallel depending on their function. This allows the fascial structures to appropriately adapt to the different stretching requests received.

There are two types of fiber that are always present, and it is only the proportion that changes depending on the load request.

Collagen Fibers

Collagen is the most abundant protein in the human body and in connective tissue. It makes the fascia strong and guarantees protection against excessive stretching. It is a structural protein that constitutes the main fibrous element of skin, tendons, cartilage, bone, teeth, membranes, the cornea, and blood vessels of all vertebrates. In the skin, it is responsible for the mechanical protection of the body, while its other roles include the well-being of the dermis and other organs (albeit indirectly), prevention against skin dehydration, maintenance of tissue elasticity and firmness, and minimization of wrinkles.

Think of collagen as a triple rope. Three chains of polypeptides form the helical structure of this protein; the right-leaning and left-leaning shape of these chains increases its load capacity. Think of a rope. The more you twist the strands, the greater the tolerable load will be. Then imagine unwinding it in the opposite direction to its curves; the various strands of the rope will tend to loosen and fray, making it weaker.

The function of collagen is to resist traction forces while proteoglycans soften the compression forces. These two roles in the network explain the innovative success of the use of reinforced concrete in the construction industry. The steel, which represents the collagen in connective tissue, assumes the role of withstanding the high traction force, while the concrete, which represents the proteoglycans in the matrix, can counteract the compression. This property is called viscoelasticity. In the human body, the interaction of the two functions defines the shape and position of the organs and muscles. Collagen fiber characteristics include the following:

- They make the fascia resistant and stable.
- They normally reproduce within 10 to 18 months.
- They are flexible despite the fact that individual fibers are not very elastic.
- They are predominantly crimped.
- They are able to stretch longitudinally.
- They offer high resistance to stretching and minimal resistance to compression.

The discovery of connective tissue cells known as myofibroblasts, located between fascial collagen fibers and with contractile properties similar to smooth muscle, was particularly noteworthy. This discovery demonstrated the ability of connective tissue to contract in certain situations [29].

Elastic Fibers

Elastic fibers are primarily made up of the protein elastin. In some organs it is very important for the connective tissue to be elastic. The elastic fibers in the tissue can slide, stretch, or deform depending on external stresses, but they are able to return the tissue in which they are found to its original condition. By way of example, connective tissue rich in elastic fibers can be found in the urinary bladder and in the tunica media of the arteries. Its structure does not change much throughout life.

Elastic fiber characteristics include the following:

- They make the fascia elastic.
- They can stretch to 150 percent of their original length.
- After stretching, they easily return to their original length.
- They work best at 98.6 °F (37 °C).

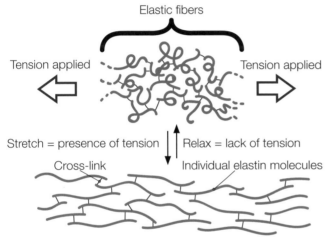

As their name suggests, elastic fibers make tissue elastic and mobile. Any stress exerted on the tendons and ligaments is first absorbed by the elastic fibers before being evenly transferred to the collagen fibers. Elastic fibers ensure that collagen fibers maintain their wavy or crimped shape.

Collaboration between elastic fibers and collagen fibers.

Reticular Fibers

Reticular fibers are made of type III collagen. These fibers tend to produce collateral fibers that are organized into net structures (which is why they are called reticular). Their role is to support the neighboring cells (blood vessels and dermis, nerves, and lymph nodes). In other words, collagen is made up of ropes, and the reticular fibers form the mesh that extends between them.

Cells

Most of the cells are fibroblasts, which are responsible for producing the ground substance, collagen, and elastin, as well as the cross-linking proteins. They are activated by **mechanical** deformation. They react to compression loads by producing proteoglycans and glycosaminoglycans, the shock absorbers of the tissue. They react to traction stress by synthesizing collagen and elastin. Therefore, the decisive stimuli to improve the quality of connective tissue by movement primarily consist of attaining the right balance between compression and traction loads. In addition, elastic movements favor the crimped, wavy arrangement of collagen, while the heat produced by the body seems to have positive effects on the fascia.

It has been discovered that the fibroblasts in the fascial sheaths of individual muscle fibers are activated by 20 to 30 percent of the maximum force, even with a very slight production of force. This makes sense because stabilizing the muscle and holding it in position is the primary role of fascial tissue.

Water

The ground substance of the fascia is made up of approximately two-thirds water, known as bound water (water that binds to the solid substances of the fascia). This bound form allows water to flow like it does on a riverbed, without dispersing in all directions, but rather acting as a means of transport similar to an aqueduct.

The water balance in the fascia is governed by osmosis, by which water consumed is removed by the fascia in order to absorb other fresh and clean water. The continuous exchange of water is important for the intake and removal of metabolic products and for the immune cells. Think of a pond with little or no water exchange: Over time it becomes putrid and starts to smell. In contrast, a lake with constant water exchange stays clean and fresh.

OSMOSIS

Osmosis is a phenomenon by which water flows between two solutions separated by a semipermeable membrane. The phenomenon is generally caused by a difference in concentrations, where the water flows from the solution with the lowest concentration to the solution with the highest concentration.

By applying external pressure to a semipermeable membrane, the direction of osmotic flow reverses, meaning that the solvent flows from the more concentrated solution to the less concentrated solution. This is known as reverse osmosis [27].

Think of it as wringing out a sponge full of stagnant water so that it can soak up clean water. The water content in the interstitial space is around 70 percent.

Conclusions

We have now examined the main components of fascial tissue and their most important functions. That just leaves nerve cells (neurons), which are discussed in section 1.6. At first glance, the list of the different types of connective tissue may not seem particularly clear, but upon more detailed examination, we can identify the following key functions:

- Shape: Surrounds, supports, protects, fills, sustains, organizes
- Movement: Transfers and stores force, holds tension, stretches
- Supply: Supports the metabolism, transports fluids, nourishes
- Communication: Receives and transmits stimuli and information

Because the different functions practically always occur together, they complement and influence each other.

We can also conclude that the biomechanical properties of fascial tissue depend on the number and orientation of the collagen fibers compared to the ground substance, and on the proportion of collagen fibers compared to elastin fibers.

By combining and overlapping the main functions of the fascial structures and the FReE training method, the following **movement square** emerges.

The Movement Square

1. FEEL IT = Perception and activation
2. MOBILITY = Functional mobility
3. STRETCH = Fascia stretching and shape
4. ENERGY = Movement as elastic energy
5. RELEASE = Fascial release

The movement square shall be referred to throughout the book.

1.4. CONSIDERATIONS AND STUDIES

Having analyzed fascial tissue, we can draw our conclusions from interesting studies that bring clarity to the work we do in practice.

This figure at bottom left from R. Schleip and D. G. Müller [60] represents the increased elastic memory capacity. Regular oscillatory exercise, such as going for a run every day, increases the tissue's ability to store elastic force. As such, it acts exactly like a spring.

The figure at bottom right from R. Schleip and D. G. Müller [60] shows the different collagen structure found in young, active people versus inactive adults. As you can see, unstimulated fascia loses its organization, and its structure becomes irregular and confused: The architecture of the collagen responds to the load.

It is interesting to note how the bidirectional collagen fibers in the fascial tissue of young people are extremely wavy and crimped (left figure) and form a reticular structure similar to compression stockings [63]. In contrast, we tend to lose elasticity as we age, and the fibers of the fascial architecture adopt a more random and multidirectional arrangement (right figure). A lack of movement and incorrect posture associated with age give rise to disorganized and random interconnections in the fascia. The fibers lose their elasticity and do not slide over one other; rather, they stick to each other, form tissue adhesions and, in the worst case, become matted [31].

The aim of fascia training is to stimulate the fibroblasts through strategies that use elastic energy and dynamic stretching in order to maintain a youthful and elastic fascial architecture. This is achieved by multidirectional stretching to stimulate the elasticity of the tissue [41].

For technicians, it is very important to bear in mind the following: People who are constantly active stimulate and improve the structural organization of their fascia, while sedentary people who get up off the sofa or leave their hospital bed to get back to exercising face two challenges: 1) remodeling the spiral reticular structure and 2) reshaping the crimps. Both challenges require more time than simple muscle formation.

In 1996, Staubesand found that the lattice of the fascia of young women was more bidirectional than in older women [63].

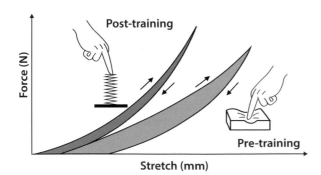

Reprinted by permission from R. Schleip and D.G. Müller, "Training Principles for Fascial Connective Tissues: Scientific Foundation and Suggested Practical Applications," *Journal of Bodywork and Movement Therapies* 17, no. 1 (2013): 1-13.

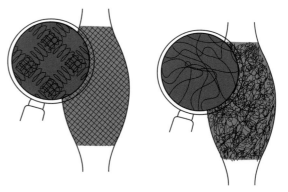

Reprinted by permission from R. Schleip and D.G. Müller, "Training Principles for Fascial Connective Tissues: Scientific Foundation and Suggested Practical Applications," *Journal of Bodywork and Movement Therapies* 17, no. 1 (2013): 1-13.

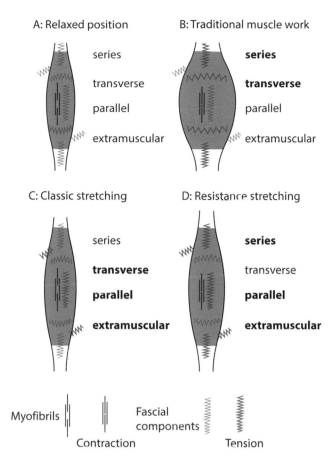

A: Relaxed position

- series
- transverse
- parallel
- extramuscular

B: Traditional muscle work

- **series**
- **transverse**
- parallel
- extramuscular

C: Classic stretching

- series
- **transverse**
- **parallel**
- **extramuscular**

D: Resistance stretching

- **series**
- transverse
- **parallel**
- **extramuscular**

Myofibrils | Fascial components | Contraction | Tension

Reprinted by permission from R. Schleip and D.G. Müller, "Training Principles for Fascial Connective Tissues: Scientific Foundation and Suggested Practical Applications," *Journal of Bodywork and Movement Therapies* 17, no. 1 (2013): 1-13.

In 2002, Jarvinen proposed that a lack of movement gives rise to the multidirectional distribution of collagen and fewer crimps [31]. In 1988, Wood found that guinea pigs that run every day produce more crimped collagen [79].

The fascia inside a muscle may be arranged in series, diagonally, parallel, or outside the muscle. This figure above shows the load of the different fascial components (taken from R. Schleip and D. G. Müller [60]).

A. **Relaxed position**: The muscle fibers are relaxed, and the length of the muscle is normal. None of the fascial elements are stretched.

B. **Traditional muscle work**: The muscle fibers are contracted. The fascial tissues are tensed, both those arranged in series with the myofibers as well as those arranged diagonally to the muscle fibers.

C. **Classic stretching**: The muscle fibers are relaxed, and the muscle is stretched. The fascial tissue arranged in parallel to the muscle fibers is tensed; however, the fascial tissue arranged in series with the myofibers is not sufficiently tensed.

D. **Resistance stretching**: The muscle fibers are tensed; most of the fascial components are stretched and stimulated.

1.5. COLLAGEN TURNOVER

The fascial network undergoes a continuous process of catabolism and reconstruction. Half of the collagen has reproduced after about one year. The more stuck or matted the tissue is, the longer it will take to regenerate. The body's power to reconstruct and regenerate depends on its ability to transform its own resources from one to another. Protein synthesis is a metabolic process that requires the presence of collagen as a catalyst. As the amount of collagen in the body falls, so too do the resources and the catalyst. Various factors could damage and at the same time reduce collagen production in the body, including the following:

- Poor nutrition
- Sedentary lifestyle
- Alcohol abuse

The body produces collagen every day but, because production decreases with age, the available collagen supply rapidly ceases to be sufficient, and the various parts of the body start to gradually deteriorate. The first obvious signs are the appearance of wrinkles and joint pain, but these are just some of the effects of the main problem. Given that health comes from within the body, the secret to looking young and healthy is to deal with the source of degeneration before the process begins.

How long should you wait between one myofascial training session and the next? To answer this question, let's refer once again to the studies conducted.

Collagen Turnover After Exercise

The concept of supercompensation used in muscle training also applies to the fascia. It is essential to combine training and rest to improve performance and strengthen the body.

Research by Magnusson et al., 2008, shows that collagen synthesis in the tendons increases after exercise. However, the collagen degradation rate by stimulated fibroblasts also increases. In fact, it is interesting to note that on days 1 and 2 after physical exercise, the collagen degradation rate is higher than the collagen synthesis rate, while from day 3 onward this situation is reversed. Regeneration is complete after 72 hours, with superior quality compared to baseline. As a result, a break of 2 or 3 days is needed to respect this natural process; another training session during this time would make little sense.

1.6. THE NERVOUS SYSTEM: INTRODUCTION

We cannot talk about the fascia without mentioning the nervous system. We know that it is a highly complex subject, and I don't intend to go into too much detail or use complicated jargon. However, a basic knowledge is essential in order to improve and to be able to confidently propose a comprehensive training program.

Let's start by defining the nervous system as a complex network of nerves and cells that transport messages to and from the brain and the spinal cord to different parts of the body.

It consists of both the **central** and **peripheral** nervous systems. The central nervous system (CNS) is made up of the brain, cerebellum, and spinal cord, whereas the peripheral nervous system consists of two parts:

1) Somatic nervous system
2) Autonomic nervous system

1.6.1. The Somatic Nervous System

The somatic nervous system consists of peripheral nerve fibers that carry sensory information or sensations from the peripheral organs (muscles, limbs, fascia, joints) to the central nervous system. For example, when you touch something hot, the sensory nerves carry the information about the heat to the brain, which in turn, through motor nerves, sends a message to the muscles of your hand to immediately remove it. It all takes less than a second.

1.6.2. The Autonomic Nervous System

The autonomic nervous system (ANS), also called the vegetative or visceral nervous system, together with cells and fibers, innervates the internal organs and glands, controlling the so-called vegetative functions, which are generally involuntary. For this reason it is also known as the **involuntary autonomic system**. The ANS is part of the peripheral nervous system. It is responsible for regulating the body's homeostasis (equilibrium) and is an involuntary nerve and motor system that operates by autonomic mechanisms in the form of peripheral reflexes subject to CNS control.

The autonomic nervous system is divided into three sections:

1) The **sympathetic** (or **orthosympathetic**) nervous system
2) The **parasympathetic** nervous system
3) The **enteric** nervous system

The sympathetic nervous system controls fight or flight reactions, while the parasympathetic nervous system is responsible for rest and assimilation. You can think of the sympathetic nervous system as the accelerator (gas pedal) of a car and the parasympathetic nervous system as the brake.

Sympathetic Nervous System

As soon as we activate the sympathetic nervous system we are ready for action, and our heart rate, blood pressure, muscle activity, and breathing all increase.

The sympathetic nervous system performs many functions, all associated with the fight or flight reaction. Strictly speaking, the sympathetic nervous system is responsible for

- circulating catecholamines from the adrenal medulla,
- dilating the pupils by contracting the pupillary dilator muscles,
- relaxing the ciliary muscles and accommodating for long-distance vision,
- increasing the systolic volume of the heart, heart rate, and blood pressure,
- dilating the bronchi,
- dilating the coronary arteries,
- dilating the vessels of skeletal muscles,
- contracting the peripheral blood vessels,
- constricting the vessels of the skin and organs (except the heart and lungs),
- increasing hydrochloric acid synthesis by the parietal cells of the gastric glands and reducing the mobility and stimulation of the stomach sphincters,
- promoting the hydrolysis of glycogen,
- decreasing diuresis, and
- decreasing intestinal mobility with movements of the gastrointestinal tract.

Parasympathetic Nervous System

When we activate the parasympathetic system it means we are in a recovery phase: Heart rhythm slows down, muscles relax, and breathing rate slows together with the body's other activities. This system is responsible for

- constricting the pupil (miosis),
- normal functioning of the tear glands,
- abundant secretion of the salivary glands,

Parasympathetic System

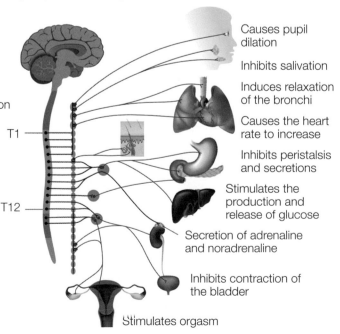

Sympathetic System

- contracting the smooth muscle of the lungs, reducing air intake volume,
- dilating the blood vessels of the genitals and glands of the digestive system,
- reducing the systolic volume of the heart, heart rate, and blood pressure,
- constricting the coronary arteries,
- increasing secretion of the stomach, inhibiting the sphincters, and increasing mobility,
- greater mobility of the intestinal walls,
- the liver promoting glycogenesis and increasing bile secretion,
- the pancreas promoting secretion, and
- the bladder stimulating the wall and inhibiting the sphincter.

Parasympathetic innervation "prevails over sympathetic innervation in the salivary glands, tear glands, and in erectile tissue (cavernous and spongy bodies of the penis and clitoris)" [73].

Our longevity depends on the continual equilibrium between the sympathetic and parasympathetic systems. If we always live life at full throttle, sooner or later we will fall apart. We should be aware of these mechanisms to improve our lives, our training, our recovery, and our performance. These mechanisms are, and must be, part of our training and of our lives.

Enteric Nervous System

This is the third part of the autonomic nervous system. The enteric nervous system is a complex network of nerve fibers that innervate the abdominal organs like the gastrointestinal tract, pancreas, and gallbladder. It contains almost 100 million nerves.

1.7. THE BRAIN

We are now going to enter the wonderful world of the brain. We know that it is a **plastic** organ; it is not only a decision-maker, but it is also able to listen and change. This **ability to change** brain connections, and therefore function, is one of the most interesting aspects that can be influenced by movement because it means that we can change the brain.

Functions of the Brain Lobes

- Frontal: Reasoning, planning, emotions, voluntary movements, language

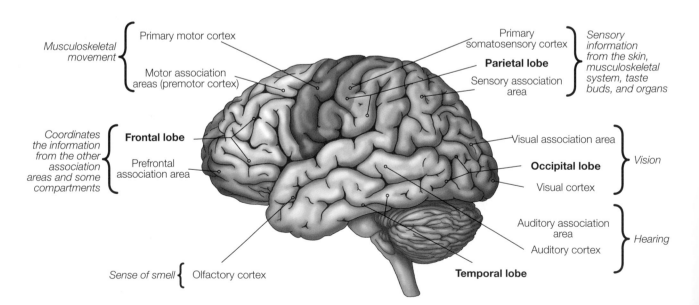

- Temporal: Sound volume and frequency, memory training and consolidation
- Parietal: Sensory perceptions, pain, hearing, sight
- Occipital: Decodes visual information to identify objects

The volume of the brain decreases with age: It starts to shrink from the age of 25 and is 15 percent smaller by the age of 40, while at 50, cell death intensifies and cognitive ability falls by 30 percent.

The brain has inexhaustible resources and, if properly stimulated, can create new neural circuits, which are the only resource available to slow deterioration of brain function.

If we compare our brain to a computer that is significantly more advanced than any modern machine, so far we have only focused on the "hardware". It is now time to consider the "software" of the brain: the tasks that it performs. All the different areas of the brain must be activated to keep them awake and active;

otherwise, the synapses will deteriorate over time as per the *use it or lose it* principle. Our brain learns new notions and new tasks and modifies existing files (brain maps) through a process known as neuroplasticity.

1.8. PERCEPTION AND RECEPTORS

The fascial network is our body's largest and densest sensory organ. Eighty percent of all free nerve endings/receptors are found in the superficial fascia.

The fascia's receptor properties are well established thanks to the presence of various nerve receptors (Ruffini, Pacini), which are sufficient in number to induce proprioceptor changes against specific problems, such as lower back pain, as well as to provide feedback on movement [75].

The fascial continuum throughout the body, the mechanical role, and the ability of fibroblasts to communicate with each other through junctions suggest that fascia may

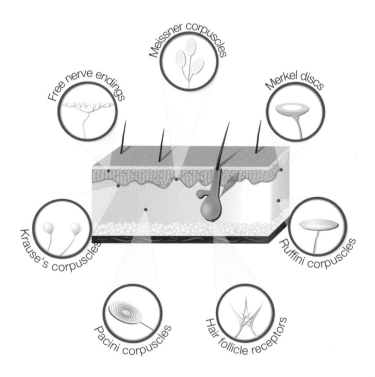

act as a mechanical-sensory signaling system, with an integrative function similar to that of the nervous system [36; 35]. It is fundamental to the tensegrity of the entire system.

Our brain needs information, which it receives through **receptors**—nerve stations that perceive certain stimuli and send the signal to the central nervous system. For simplicity, I call them "sentries" or "monitors" that monitor what is happening on the periphery. The receptors are highly specialized at detecting very particular stimuli.

Numerous mechanoreceptors are found within and around the muscles, joints, tendons, ligaments, and capsules, etc. **Mechanoreceptors** are the fastest and most powerful of the proprioceptors and transport a huge amount of information concerning the status of the periphery to the brain. The mechanoreceptors (sensory receptors) transmit information pertaining to position or movement.

Nociceptors (pain receptors) can be mechanical or thermal and are responsible for pain perception. These types of cells can change or inhibit one another. Movement may inhibit pain, while a lack of movement may increase it.

The most recent studies conducted by Schleip [61] identified six times more receptors in the fascia around each individual muscle than in the muscle itself. The muscle itself, by comparison, is rather insensitive (apart from a few exceptions: suboccipital muscles, eye muscles, and plantar muscles). An analysis of the fascial microstructure reveals that the fascia has a life of its own, with the ability to develop its own reactions and its own movements thanks to an abundant network of nerves, as well as multiple smooth muscle cells [61].

The fascial system comprises a dense population of mechanoreceptors, such as Golgi receptors, which are not only found in tendons (10 percent) but also in ligaments, joint capsules, and muscle-tendon junctions (90 percent) [61].

We will now consider the different types of mechanoreceptor in detail.

1.8.1. Mechanoreceptors

Mechanoreceptors are sensory receptors that are stimulated by a change to their shape. They provide information about the shape, consistency, and relationship of surrounding objects. They can be classified into the following four groups [61].

Golgi   Pacini   Ruffini   Free nerve endings

Golgi Receptors
Golgi receptors react to a vertical contraction; they are force receptors, and because they are arranged in series to the muscle, they respond to changes in force that occur in the tendon heads.

Where Are They Found?
- In the muscle-tendon system
- In the epimysium (around the muscle)
- In the aponeuroses
- In the peripheral ligaments of the joints
- In the capsules

Pacini Corpuscles
They are sensitive to fast pressure changes, vibrations, swaying, and sudden quick movements (see the friction exercises with the foam roller in section 4.3.3). They constantly require new stimuli to keep active, and they react within two seconds.

Where Are They Found?

- In the myotendinous junctions
- In the deep layers of joint capsules
- In the spinal ligaments (ligaments of the back)
- In the muscle fascia

Ruffini Corpuscles

They react to slow impulse changes and to prolonged pressure. Stimulus on the large areas of soft tissue has a calming effect on the sympathetic system, reducing its activity (Berg e Capri, 1999). This explains the deeply relaxing effect of these techniques. Ruffini corpuscles are primarily activated by applying tangential and transverse forces [61].

A large collection of Ruffini receptors is located in the thoracolumbar fascia.

Breathing has an excellent relaxation effect and helps to reduce tone (see "Release" in section 8.4).

Where Are They Found?

- In all types of fascial tissue
- In the outer layers of joint capsules
- In the aponeuroses
- In the ligaments

Free Nerve Endings

These may be the least known of all the receptors. The free nerve endings of type III (myelinated) and IV (unmyelinated) nerve fibers are sensory receptors that are found in large numbers. They transmit sensory information from the myofascial system to the CNS. They are called interstitial muscle receptors, 10 percent of which are type III and 90 percent are type IV, and they are the most versatile to changes. They respond to pressure and to mechanical tension [61].

Where Are They Found?

- Almost everywhere, including bones
- In the periosteum (skin of the bone), which has a particular abundance of nerve endings
- In adipose tissue
- In the superficial fascia

1.9. CONCLUSIONS

In this chapter we have seen how the sensory organs inform the brain of everything that is happening around us through receptors, special neurons that react to stimuli by producing nerve impulses. When the brain receives these impulses, it translates them into sensations.

The brain is interested in what is happening interstitially in the fascia. In addition to the vestibular system and the numerous skin sensors, fascial sensors are absolutely vital in order to know what is happening to our body in space.

Ignoring our sensory messages (*no pain, no gain*) is a sure-fire way to cause short- or long-term fascia injury. In contrast, developing a balanced sense of proprioception and kinesthesia will help to prolong our body's capabilities into older age. This explains why it is important to stimulate our nervous system in different ways.

"When you deal with the fascia you deal and do business with the branch offices of the brain."

Andrew Taylor Still
founder of osteopathy, 1899

MYOFASCIAL LINES

2.1. THE MYOFASCIAL MERIDIANS OF THOMAS MYERS

Thomas Myers discovered that synergistically coordinated myofascial connections are spread longitudinally across the whole body. These create a map that helps us to read and interpret our body and explain how movement and force are distributed within the body as a whole. Myers called these lines **myofascial meridians**.

This organized road network was discovered by in-depth post mortem examinations and research on connective tissue. These interconnected roads wrap the body in a three-dimensional webbing and functionally support the body's primary movements and maintain posture. In practice, force, tension, compensations, and most spatial movements are distributed along these lines.

This discovery has changed our understanding of the body and given rise to new educational approaches. From a structural perspective, the three-dimensional musculoskeletal anatomy is a network that balances the distribution of forces in our body.

The logic of these lines is an integral part of all my training programs because I can visualize them, find them, feel them, and use them. All this helps to improve the training itself because by using a few simple tricks, the lines can be activated to better distribute force.

A continuous line of tension
A continuous plane of fascia
A series of interconnected myofascial units creating a VOLUME

2.2. LANGUAGE OF THE BOOK

Before going into more detail, I would like to say a few words about the language I use and how this chapter is divided.

FIND IT (FIND THE LINE)
This section will help you to orient yourself according to the map of the myofascial lines.

FEEL IT
Having used the map, you will be able to link the points together. This will help your mind to perceive the relationship between your muscles and the myofascial lines and to feel the connections and muscle contractions. You will learn to explore the lines in detail.

USE IT
Finally, in this last section you will explore the myofascial lines in motion.

For simplicity, I use the following abbreviations throughout the book:
SBL = Superficial Back Line
SFL = Superficial Front Line
LL = Lateral Line
DFL = Deep Front Line
SL = Spiral Line
AL = Arm Lines
DFAL = Deep Front Arm Line
SFAL = Superficial Front Arm Line
DBAL = Deep Back Arm Line
SBAL – Superficial Back Arm Line
FL = Functional Line

2.3. SUPERFICIAL BACK LINE (SBL)

The Superficial Back Line is a cardinal main line that connects the entire surface of the back of the body from the bottom of the foot to the top of the head in two parts: foot–knee, and knee–forehead. The SBL connects, stabilizes, and moves the back of the body from the toes to the hollow of the knee and from the knee to the eyebrows.

Standing up with your legs straight, the SBL acts as a continuous myofascial line.

FIND IT

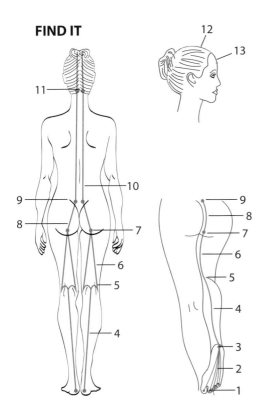

Bony stations		Myofascial tracks
Frontal bone, supraorbital ridge	13	
	12	Galea aponeurotica/ epicranial fascia
Occipital crest	11	
	10	Sacrolumbar fascia/spinal erectors
Sacrum	9	
	8	Sacrotuberous ligament
Ischial tuberosity	7	
	6	Hamstrings
Condyle of the femur	5	
	4	Gastrocnemius/Achilles tendon
Calcaneus	3	
	2	Plantar fascia and short toe flexors
Plantar surface of the phalanges of the toes	1	

Posture and Movement

The main effect of the SBL is on postural and motor patterns in the sagittal plane. The primary actions of the SBL are as follows:

- Extension of the head
- Extension of the spinal column
- Flexion of the sacrum (nutation)
- Extension of the pelvis
- Flexion of the knees
- Plantar flexion of the ankles

At the same time, it prevents involuntary movements or excessive movements in the opposite direction.

Because we are born in a flexed (fetal) position, the development of strength, capacity, and balance in the SBL is associated with slow evolutionary movement that from this primary flexion, matures into complete and lasting extension.

The high concentration of slow muscle fibers (red fibers) enable the SBL to work against gravity with enormous efficiency.

In the exercise chapters, when you are instructed to adopt a particular position I will sometimes refer to the upper part and lower part of the SBL (lower part meaning the lower limbs and upper part meaning the chest), because the whole line cannot always be activated in the same way.

Achilles Tendon and Calcaneus (Heel Bone)

The Achilles tendon and calcaneus require special mention. The Achilles tendon connects the calf muscle (gastrocnemius and soleus muscles) to the heel bone to improve the fluidity of movement of the ankle and knee. Think of the heel as the kneecap of the ankle.

Around the calcaneus there is actually a strong fascial network that connects the plantar fascia to the Achilles tendon. The calcaneus separates the elastic tissue of the Achilles tendon from the ankle.

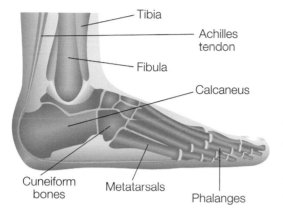

The plantar fascia and the fascia associated with the Achilles tendon act like the string of a bow, with the heel as the arrow. When the string of the bow is subjected to continuous and excessive hypertension, it may push the heel in front of the subtalar joint of the ankle. This displacement has a negative effect on the supporting base.

On a purely empirical basis, Thomas Myers found that the proportion of the hindfoot and forefoot should be one-third or one-quarter (between heel and malleolus) and two-thirds or three-quarters from the malleolus to the fifth metatarsal head to offer effective support. Without this balanced support the rest of the body becomes destabilized.

The exercises you will find in the following chapters shall refer to this bow, which acts as a catapult.

FEEL IT

Explore the SBL in minute detail: Connect the points as you adopt the positions. (The exercises that I propose here are purely for illustrative purposes: They shall be covered specifically in the exercise chapters).

Roll down: Lean forward and you should feel the SBL activate and stretch from the soles of your feet to your head.

Extension: Raise your arms above your head and move your pelvis slightly forward, making the shape of an arch. You should feel the SBL activate and strengthen.

USE IT

Explore the SBL in motion. Feel the strength of the SBL, and how it stretches and slides, as well as how these vary between the different points of the SBL.

Stretching the calf muscles and Achilles tendon.

Standing upright.

Stretching the hamstrings.

Stretching the lumbar spinal erectors and lumbar fascia.

Isometric strengthening from your toes to your head.

Stretching the spinal erectors.

Compensations in the SBL

The following postural patterns are often associated with the SBL:

Stretching the plantar fascia and short flexors.

- Limitation of the ankles in plantar flexion
- Hyperextension of the knees (sliding of the femur forward over the tibia)

- Tight hamstrings
- Forward positioning of the pelvis (ventral)
- Nutation of the sacrum
- Hyperextension of the thoracic spine (short thoracic extensors)
- Thoracic area: Myofascial tissue stretched and enlarged = thoracic flexion
- Forward positioning and rotation of the atlanto-occipital joint (ventral)
- Interruption of movement of the eye–spine–connection

This last bullet point highlights the fascinating relationship between the eyes and the spinal column.

Note: I have listed compensation patterns on the sagittal plane, but we can also create compensations on the right or left side of the SBL. The importance of this is not to be underestimated.

In the early 1980s, J. Hegge developed a series of techniques to improve the part of human movement controlled by vision. He describes in particular how these techniques can be used to reduce excess muscle tension in the neck and shoulders, as well as to improve neck movement [26].

Superficial Back Line and Training
SBL training should encompass the following exercises:

- Eccentric strength training in all layers
- Isometric and concentric strength training
- Resistance strength training (tonic line), slow-twitch muscle fibers
- Elastic strength training (expansion of the elastic Achilles tendon, close to the head it stabilizes)
- Differentiated flexibility
- Sliding in and around the SBL
- Openness from head to toe (not sticky)

Finally with regard to training, it is important to note that all the lines influence each other and that the LL and the back arm lines require attention in relation to training of the SBL.

Focus On
With regard to the SBL, the following points should be taken into consideration:

- The main function of the SBL is to support an upright and open posture.
- This line contains more slow-twitch muscle fibers than the SFL.
- It supports us in full extension and is resistant to bending in flexion.
- It lifts us up and keeps us upright.
- Its main movements are on the sagittal plane.
- With the exception of the knee and the plantar flexion of the ankle, its function is creating extension.
- Imbalances in the SBL tend to have an impact on the secondary curves.
- The suboccipital muscles can be thought of as the cornerstone of the SBL.
- The SBL and SFL initiate rotation when walking.
- The fascia flows from high to low.

Please see the specific exercises in later chapters that will help you to make and keep the SBL long, strong, lubricated, and adaptable.

2.4. SUPERFICIAL FRONT LINE (SFL)

The Superficial Front Line is a cardinal main line that connects the entire front surface of the body from the toes to the sides of the skull in two parts – from the toes to the pelvis and from the pelvis to the head. When they are fully extended, like when standing upright, they act as a continuous integrated myofascial line.

FIND IT

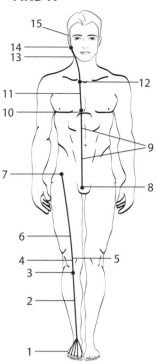

Bony stations		Myofascial tracks
	15	Fascia of the scalp
Mastoid process	14	
	13	Sternocleidomastoid
Manubrium of the sternum	12	
	11	Sternal/sternochondral fascia
Fifth rib	10	
	9	Rectus abdominis
Pubic tubercle	8	
Anterior inferior iliac spine	7	
	6	Rectus femoris/vastus lateralis, medialis, intermedius
Patella	5	
	4	Patellar tendon
Tibial tuberosity	3	
	2	Short and long extensors of the toes/tibialis anterior/anterior crural compartment
Dorsal surface of the phalanges	1	

Posture and Movement

The SFL works together with and balances the SBL, influencing postural and motor patterns in the sagittal plane. The primary actions of the SFL are as follows:

- Extension of the upper part of the cervical spine, flexion of the lower part of the cervical spine
- Flexion of the thoracic and lumbar spine
- Flexion of the hip
- Extension of the knee
- Thoracic extension and stabilization of the ankle (particularly supporting the heel)
- Supination of the foot

Proportionally, the SFL contains more fast-twitch muscle fibers (white fibers) than the SBL. The muscles of the SFL contract quickly and with enormous strength, similar to how you react to something that scares you suddenly.

FEEL IT

Explore the SFL in minute detail. Connect the points as you adopt the positions. (The exercises here are purely for illustrative purposes: They shall be covered in more detail in the exercise chapters).

Extension: Raise your arms above your head and stretch your chest. Move your pelvis forward, making the shape of an arch. You should feel the entire front of your body activate and stretch, from your feet to the top of your head.

Plank position (Front stabilization): Feel the SFL activate and strengthen.

Strengthening of the quadriceps.

USE IT
Explore the SFL in motion. Feel the strength of the SFL, and how it stretches and slides, as well as how these vary between the different points of the SFL.

Stretching of the tibial tract.

Stretching of the rectus abdominis and sternocleidomastoid.

Compensations in the SFL
The following postural patterns are often associated with the SFL:

- Limitation of the ankles in plantar flexion
- Hyperextension of the knees
- Anteversion of the pelvis
- Anterior limitation of breathing and the ribs
- Head position forward/tilted downward

Note: I have listed compensation patterns on the sagittal plane, but we can also create imbalances on the right or left side. The importance of this is not to be underestimated.

Stretching of the quadriceps.

Superficial Front Line and Training
SFL training should encompass the following exercises:

- Eccentric strength training
- Differentiated isometric and concentric strength training

- Elastic strength training
- Functional flexibility training
- Sliding in and around the SFL
- Balance of training for both the upper body and lower body

One final consideration: If you focus on the SFL during training, it is also particularly important to consider the function of the DFL. To successfully balance the SFL, it is very important to balance the front arm lines.

Focus On

With regard to the SFL, the following points should be taken into consideration:

- The SFL contains more fast-twitch muscle fibers than the SBL.
- It is reactive.
- It balances the SBL.
- Its main movements are on the sagittal plane.
- Its primary function is flexion, with the exception of the extension of the knees, the foot, and the upper cervical spine.
- The fascia flows from low to high.
- It provides tensile support from above to lift those parts of the skeleton that cross the line of gravity: pubis, rib cage, and face.
- It protects sensitive parts of the body: internal and sexual organs.

Please see the specific exercises in later chapters that will help you to make and keep the SFL long, strong, relaxed, and adaptable.

In the exercise chapters, when you are instructed to adopt a particular position, I will sometimes refer to the upper part and lower part of the SFL (lower part meaning the lower limbs and upper part meaning the chest), because the whole line cannot always be activated in the same way.

2.5. LATERAL LINE (LL)

The Lateral Line (LL) crosses each side of the body, starting from the medial and lateral point of the middle of the foot, passing around and outside the ankle, climbing up the lateral tract of the leg and thigh, and intertwining along the trunk before arriving at the skull near the ear.

FIND IT

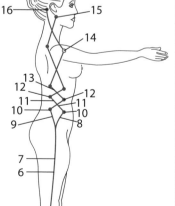

Bony stations		Myofascial tracks
Occipital margin/mastoid process	16	
	15	Splenius capitis/sternocleidomastoid
First and second ribs	14	
	13	External and internal intercostal muscles
Ribs	12	
	11	External obliques
ASIS, PSIS, and iliac crest	10	
	9	Gluteus maximus
	8	Tensor fascia latae
	7	Iliotibial tract
	6	Abductor muscles
Lateral condyle of tibia	5	
	4	Anterior ligament of the head of the fibula
Head of the fibula	3	
	2	Fibular (peroneal) muscles, lateral crural compartment
Base of first and fifth metatarsals	1	

Posture and Movement

The LL is important for both posture and movement. It balances the right and left sides of the body. It transmits forces to the other myofascial lines. Although the LL primarily acts on the coronal (frontal) plane, its movements also include the spirals and counter-rotations of the chest.

The LL helps to create the body's lateral curve – lateral flexion of the trunk, abduction of the hip, and eversion of the foot – but also acts as an adjustable "brake" for lateral movements and trunk rotations. Its primary function is to stabilize the body in motion. The primary actions of the LL are as follows:

- Eversion, pronation, and plantar flexion of the foot
- Abduction of the hip
- Lateral flexion of the spinal column and head
- Rotation of the spinal column and ribs

FEEL IT

Explore the LL in minute detail. Connect the points as you adopt the positions. (The exercises here are purely for illustrative purposes: They shall be covered in greater detail in the exercise chapters).

Standing upright: Place your right foot behind your left and raise your right arm above your head, gently stretching and tilting your chest to the left. You should feel the LL in your right side activate and stretch from the outside of your foot to the pinkie of your hand.

Place your hands on the wall and raise your left leg. You should feel both sides of your chest and legs activate and strengthen.

USE IT

Explore the LL in motion. Feel the strength of the LL, and how it stretches and slides, as well as how these vary between the different points of the LL.

Stretching of the entire LL.

Stretching and sliding of the upper part.

Strengthening of the abductors, external obliques, and intercostals.

Strengthening of the abductors, strengthening and stretching of the external obliques and intercostals.

Stretching of the entire LL.

45

Compensations in the LL

The following postural patterns are often associated with the LL:

- Limitation or excessive pronation or supination of the ankle
- Limitation in thoracic extension
- Varus or valgus knee
- Limitation of adduction/chronic tension in the abductors
- Lumbar compression
- Lateral bending of the spinal column
- Misalignment of the chest in relation to the pelvis
- Reduced space between sternum and coccyx
- Limitation in the shoulders due to excessive participation in stabilizing the head

Lateral Line and Training

LL training should encompass the following exercises:

- Dynamic stability
- Eccentric, isometric, and concentric strength training
- Elastic strength training
- Functional flexibility training
- Sliding in and around the LL
- Expandable, requires space
- Rotation movements in the chest

In terms of the relationship between the LL and the DFL, it should be noted that in the lower body, the LL and the DFL are balanced in many ways, while in the upper body they are connected through breathing.

Finally, the myofascial axial structures of the LL (in the chest) are intertwined with the Arm Lines.

Focus On

With regard to the LL, the following points should be taken into consideration:

- Its primary function is to balance the left and right sides and the SFL/SBL.
- It also mediates the forces between the other lines (AL, SL).
- It transfers forces between the superficial lines.
- It connects with the SBL and SFL in the chest.
- The upper part is closely connected to the Arm Lines and works independently to stabilize the head laterally.
- The primary movements of the LL are eversion of the foot, abduction of the hip, lateral flexion, and rotation of the spinal column.
- It acts as an adjustable "brake" for lateral movements and rotations of the trunk.
- The LL is a sensory line.

In some subjects, particularly explosive athletes who are capable of great acceleration, the gluteal muscles and the tensor fasciae latae are extremely strong, rigid, and tight, which forces the adductors to lose tone and to constantly act on excessive lengths.

Let me explain the concept: A muscle, in addition to its functional length, can be shortened due to excessive tone, thickening of the muscle fascia, fascial adhesions, etc. However, that same muscle could remain locked in a stretched position, that is, be forced to maintain a more elongated shape. If the agonist is shortened, or rather too active, the antagonist must yield tension and position itself in a state that allows the agonist to work. Its primary function is control, and if it has excessive eccentric tension it can become fibrotic over time. Therefore, the muscle will be locked in a stretched position to maintain the non-physiological lengths.

Please see the exercises in later chapters that will help you to make and keep the LL long, strong, relaxed, and adaptable.

In the exercise chapters, I will sometimes refer to the upper part and lower part of the LL (lower part meaning the lower limbs and upper part meaning the chest), because the whole line cannot always be activated in the same way.

2.6. DEEP FRONT LINE (DFL)

Let's now take a look at the DFL. We are discussing it right after the LL because its function is to promote efficient breathing and to improve posture and movement. The DFL is our nucleus, our center, our core that makes us grow from the inside outward.

Framed by the cardinal lines and surrounded by the Spiral Line and the Functional Line, the DFL creates the myofascial core of the body upon which the proper functioning of all the other meridians depends. It begins deep in the sole of the foot, like a hand that wraps around the plantar arch, and rises by passing just behind the bones of the leg. From the back of the knee, it continues its course through the thigh and in front of the hip joint, the pelvis, and the lumbar spine. It fills much of the chest, expands in the rib cage, and continues upward through the neck and the nape to the mandible and the side of the head.

By comparing this line with the others, it can be seen that it should be defined in a three-dimensional space, rather than on a simple line.

FIND IT

Bony stations		Myofascial tracks
SECTION: SUPERIOR POSTERIOR **From the lumbar region to the base of the occiput**		
Basioccipital	13	
	12	Anterior longitudinal ligament of the neck and head
Lumbar vertebral bodies, transverse processes (TP)	11	
SECTION: INFERIOR COMMON AND INFERIOR POSTERIOR **From the foot to the lumbar region**		
Lumbar vertebral bodies	11	
	10	Anterior sacral fascia, anterior longitudinal ligament
Coccyx	9	
	8	Fascia of the pelvic floor, levator ani, fascia of the obturator internus
Ischial ramus	7	
	6	Posterior intermuscular septum, adductor longus and adductor brevis
Medial epicondyle of the femur	5	
	4	Popliteal fascia, capsule of the knee
Upper/lower part of the tibia/fibula	3	
	2	Tibialis posterior, long toe flexor
Plantar tarsal bones, planter surface of the toes	1	

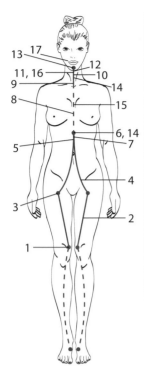

Bony stations		Myofascial tracks
SECTION: SUPERIOR CENTRAL **From the lumbar region to the base of the occiput**		
Basioccipital, cervical transverse processes	17	
	16	Prevertebral fascia, pharyngeal raphe, scalene muscles, fascia of the middle scalene
	15	Pericardium, mediastinum, parietal pleura
	14	Posterior diaphragm, pillars of the diaphragm, central tendon
Lumbar vertebral bodies	5	
SECTION: SUPERIOR ANTERIOR **From the lumbar region to the mandible**		
Mandible	13	
	12	Suprahyoid muscles
Hyoid bone	11	
	10	Infrahyoid muscles, pretracheal fascia
Posterior manubrium	9	
	8	Endothoracic fascia, transversus thoracis muscle
Posterior surface of the subcostal, cartilage, xiphoid process	7	
	6	Anterior diaphragm
Lumbar vertebral bodies	5	
SECTION: INFERIOR ANTERIOR **From the inside of the knee to the lumbar region**		
Lumbar vertebral bodies	5	
	4	Psoas, iliacus, pectineus, femoral triangle
Lesser trochanter of the femur	3	
	2	Anterior intermuscular septum, adductor brevis, adductor longus
Linea aspera of the femur, medial epicondyle of the femur	1	

Posture and Movement

The DFL supports, sustains, and straightens the body from the inside, which is why it is also called the deep core. This effortless straightening represents the DFL's unique quality: It cannot be replicated by the other superficial lines.

In terms of movement, the DFL balances the interaction between all the myofascial meridians and makes them flexible and efficient. The primary actions of the DFL are as follows:

- Lifting and dynamic stabilization of the inner arch of the foot
- Dynamic stabilization of the knee and hip
- Anterior support and straightening of the spinal column
- Dynamic stabilization of the lumbar vertebrae
- Dynamic stabilization of the chest, maintaining three-dimensional expansion and relaxation during breathing

- Balance and dynamic stabilization of the neck and head
- Counter-balance of the LL

The myofascia of the DFL is primarily composed of slow-twitch muscle fibers that are able to bear significant stress and strain. This reflects the role that the DFL performs in maintaining stability and in the subtle changes of position of the innermost structure, enabling the structures and the most superficial lines to work effortlessly and efficiently with the skeleton.

T12

We will now take a detailed look at a crucial point for the DFL: the thoracolumbar connection represented by T12 (12th thoracic vertebra).

As you will see, the term "T12" appears frequently in this book to refer to the meeting point between the lower and upper part of the DFL and the front part of the upper lumbar vertebrae. It is an important junction because blocks or retractions, which have a negative effect on breathing and posture, are common in this area.

By comparing this area with some eastern philosophies, we discover that this represents the Solar Plexus Chakra, or the third Chakra. The Chakra are centers of energy found in our body that govern our organic, mental, and emotional functions. The Solar Plexus Chakra is located at the height of the diaphragm, just below the sternum. This important energy center is home to what is commonly called ego the perception that we have of ourselves. This Chakra is associated with our personal strength, willpower, belief in ourselves, and self-esteem. We could therefore define it as our deepest me.

In the third Chakra we unload our tensions, fears, anger, and resentment. It is no coincidence that when we experience a negative emotion, we feel our own gastrointestinal system reacting; in other words, we experience somatic symptoms from psychological distress. The anger, fear, or stress that we experience on a daily basis compromises the correct functioning of our organs, often leading to digestive difficulties, metabolic problems, and many other issues.

Negative emotions hinder breathing, which is incomplete as a result: Because there is an energy block, exhaling is shortened. This occurs because we tend to feed the upper part of the body, (i.e., the mind). Physiologically speaking, shorter exhalation means that not all the carbon dioxide from the lungs is removed. What is more, not all the mental toxins are eliminated.

For our own well-being and to improve our sports performance, it is important to fully understand this crucial point of the thoracolumbar junction, which joins the diaphragm and the psoas, the upper part and lower part. The psoas is the extension of the legs. By breathing from this area (see the Release exercises in chapter 8), we release tension. In myofascial Pilates, breathing is concentrated in front of the lumbar vertebrae to reach the diaphragm, right in front of the anterior longitudinal ligament that goes upward, opposite the vertebrae.

Try it yourself: Lying on your back, with your knees bent, try to concentrate your breathing toward T12. When breathing in, the lower ribs should expand toward the floor. Now try to concentrate your breathing in front of the sacral vertebrae, rising upward to the ribs (precisely at the diaphragm). What exactly are you doing? You are stretching the spinal column from within, creating space between the vertebrae and eliminating tension and compressions.

If we look at it in greater detail, we see that the upper end of the psoas links fascially into the pillars and other posterior insertions of the diaphragm, and together they join to the anterior longitudinal ligament that goes upward in front of the vertebrae and intervertebral discs.

As we know, the connection between the psoas and the diaphragm is a crucial point for our body for both support and functionality because it connects

- the bottom with the top of the body,
- breathing with mobility of movement, and
- assimilation with elimination.

The Spinal Column and Its Hinges of Rotation

The spinal column comprises a set of vertebrae each joined to one another. It contains 33 different vertebrae: 7 cervical, 12 thoracic, 5 lumbar, 5 sacral (which in adults is a single bone called the sacrum, without discs), and 4 coccygeal vertebrae (which make up the coccyx). Seen from the side, the spine has four physiological curves that tend to cancel each other out: cervical lordosis, thoracic kyphosis, lumbar lordosis, and sacral kyphosis. The names of the vertebrae indicate the curve in which they are found and are numbered in descending order: For example, T12 is the 12th vertebra of the thoracic spine, and L5 is the 5th vertebra of the lumbar spine. The spinal column has three degrees of freedom of movement:

- Flexion-extension (sagittal plane)
- Lateral bending (coronal plane)
- Axial rotation (transverse plane)

The spinal column consists of numerous "privileged" areas, which I define as the points of rotation and the hinges of rotation. They are "privileged" because they are the inversion points of the curves of the spinal column, the areas where the rotations of the spine on the transverse plane begin. The hinges of rotation coincide with the curves of the spinal column (cervical lordosis, thoracic kyphosis, lumbar lordosis, sacral kyphosis); the points of rotation refer to the first vertebra of a physiological curve that starts the rotation movement, which is subsequently followed by the other vertebrae above and below (rotation and counter-rotation movements). The hinges of rotation and points of rotation are detailed below.

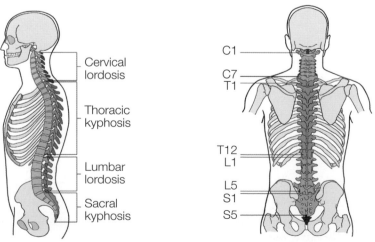

Lumbosacral joint (L5-S1). Point of rotation L5. The fulcrum for balancing the body when walking; it constitutes a "postural" risk point of the spinal column due to the structure of the S1 vertebra underneath. It is often subject to crushing, slipping, and other stress.

Thoracolumbar junction (T12-L1). T12 is the fulcrum of the thoracic hinge; Delmas defined it as a bona fide patella of the spinal column. When walking, the vertebrae from T12 up to T7 cause the trunk to rotate so as to follow the lower limb in the advancement phase. The T7 vertebrae counter-rotate as they rise.

Occipitocervical junction (C1–C2 and C7–T1). The cervical hinge contains two points of rotation: C1–C2 (atlas–axis) and C7–T1, where the dorsal hinge begins.

FEEL IT

Explore the DFL in detail. Connect the points as you adopt the positions. (The exercises here are purely for illustrative purposes. They shall be covered in greater detail in the exercise chapters).

Long toe flexors.

While standing upright, take step backward with your right leg, raise your arms above your head, and stretch and expand your chest as much as you can.

Opposing points: Push the big toe of your right foot onto the floor and push your chin forward; breathe in and out, expanding your chest in all directions. You should feel the right side of the DFL activate and stretch.

Stretching the adductors, fascia of the pelvic floor, and opening of the diaphragm.

Bridge (in descending phase); stretching of the anterior longitudinal ligament and opening of the diaphragm.

USE IT

Explore the DFL in motion. Feel the strength of the DFL, and how it stretches and slides, as well as how these vary between the different points of the DFL.

Although the DFL is not directly associated with any pertinent movement (apart from adduction of the hip), no movement is outside of its scope. The DFL can be found in almost every part of the body, surrounded or covered by other myofascia, which duplicates the roles played by the respective muscles.

Stretching and strengthening of the anterior longitudinal ligament and strengthening of the head and neck muscles.

Breathe in and out, expanding the rib cage in all directions; diaphragm.

Compensations in the DFL

The following postural patterns are often associated with the DFL:

- General shortening inside the body
- Chronic plantar flexion
- Pronation and supination
- Valgus and varus knee
- Weak pelvic floor
- Breathing difficulties
- Hyperextension or flexion of the neck
- Temporomandibular joint disorder

Deep Front Line and Training

DFL training should encompass the following exercises:

- Dynamic and static stability
- Eccentric, isometric, and concentric strength training
- Elastic strength training
- Functional flexibility training
- Sliding in and around the DFL
- Three-dimensional space, expansion

For training purposes it is important to bear in mind that balancing of the LL is vital for balance in the DFL. It is also worth noting that breathing creates space, which eliminates tension from the thoracolumbar region (T12).

Focus On

With regard to the DFL, the following points should be taken into consideration:

- It is three-dimensional: It takes up space.
- It is made up of slow-twitch muscle fibers that are able to bear significant stress and strain.
- It is the core of our body.
- It lifts us from the inside.
- It gives us stability.

In the exercise chapters, I will sometimes refer to the upper part and lower part of the DFL (lower part meaning the lower limbs and upper part meaning the chest), because the whole line cannot always be activated in the same way.

2.7. SPIRAL LINE (SL)

The Spiral Line wraps around the body like a double helix. As with the cardinal lines (SBL/SFL/LL), we have two spiral lines that wrap around the whole body from head to foot and from foot to head. Three-dimensional movement is important in sport and for everyday movements like walking.

The complete pathway of the SL of the left side begins behind the head (occipital margin) and proceeds along the neck to the region above the chest, where it crosses the spinal column and continues toward the medial border of the right scapula. At this point it plunges under the shoulder blade, continues along the ribs, and moves anteriorly, forming an intersection at the naval and reaching the left hip. From here it continues its journey downward outside the thigh, past the tibia, and under the sole of the foot. Finally, it goes back up via the posterior and external part of the leg before reaching the ischium and myofascia of the erectors, where it changes sides again (right) to go back up behind the head.

Posture and Movement

The SL helps to maintain balance between all the planes. Its function is to create and mediate spirals and rotations in the body. It also acts as an adjustable brake for excessive rotation movements. It interacts with the cardinal lines (SBL/SFL/LL) and has a close relationship with the Arm Lines, particularly the Deep Back Arm Line.

FIND IT

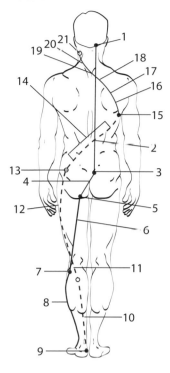

Bony stations			Myofascial tracks
Occipital margin/mastoid process/atlas and axis transverse processes	21		
		20	Splenium of the head and neck
Spinal processes (SP) of the lower cervical spine/upper thoracic spine	19		
		18	Rhomboid major and minor
Medial margin of the scapula	17		
		16	Serratus anterior
Lateral ribs	15		
		14	External oblique Abdominal aponeuroses/linea alba Internal oblique
ASIS, PSIS, and iliac crest	13		
		12	Tensor fasciae latae, iliotibial tract
Lateral condyle of the tibia	11		
		10	Anterior tibia
Base of the first metatarsal	9		
		8	Peroneus longus
Head of the fibula	7		
		6	Biceps femoris
Ischial tuberosity	5		
		4	Sacrotuberous ligament
Sacrum	3		
		2	Sacrolumbar fascia, spinal erectors
Occipital margin	1		

FEEL IT
Explore the SL with this exercise.

Standing upright, raise your outstretched arms to shoulder height, take a step forward with your right leg, bend your right elbow, and rotate your chest while looking forward. Take a step backward with your left foot, bend your left elbow, and rotate your chest. Feel the rotation

USE IT
Explore the SL in motion. You should feel the elastic and spiral movements, the force, and the sliding of the SL.

Stretching.

Dynamic stretching (recoil).

Compensations in the SL

Imbalances often occur when performing rotations, lateral movements, and lateral bending.

Spiral Line and Training

SL training should encompass the following exercises:

- Eccentric, isometric, and concentric strength training
- Elastic strength training
- Balanced flexibility
- Balance between the two spirals
- Elastic spiral movements
- Sliding in and around the SL

A further consideration for training: We should bear in mind that the SL influences and is influenced by all the other lines. Restrictions in the SL reduce the functionality of the other lines.

Focus On

With regard to the SL, the following points should be taken into consideration:

- The SL creates and mediates the body's spirals and rotations; it is the line that makes you walk.
- It maintains a close relationship with the Deep Back Arm Line.

- It promotes the equilibrium of both the body's Spiral Lines (right and left).
- It is responsible for compensation and rotation.

In the exercise chapters, when you are instructed to adopt a particular position, I will sometimes refer to the upper part and lower part of the SL (lower part meaning the lower limbs and upper part meaning the chest), because the whole line cannot always be activated in the same way.

2.8. ARM LINES (AL)

The four arm lines run from the chest to the four quadrants of the arm and to the four sides of the hand: thumb, pinkie, palm, and back of the hand. The movement of these lines is similar to that of a snake, coming to the surface before weaving their way back into the arms. They also contain more intersected myofascial connections than their corresponding leg lines. This is because the shoulders and arms of human beings are specifically designed for movement, whereas the primary role of the legs is stability. The movements that the upper limbs perform require various control and stabilization lines and therefore more connections between the lines.

FIND IT

1) Deep Front Arm Line (DFAL)

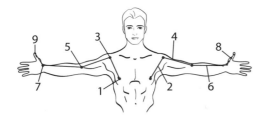

Bony stations		Myofascial tracks
Third, fourth, fifth ribs	1	
	2	Pectoralis minor, clavipectoral fascia
Coracoid process	3	
	4	Biceps brachii
Radial tuberosity	5	
	6	Radial periosteum, anterior border
Radial styloid process	7	
	8	Radial collateral ligaments, thenar muscles
Scaphoid, trapezium, exterior of the thumb	9	

2) Superficial Front Arm Line (SFAL)

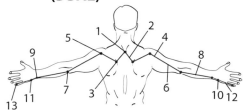

Bony stations		Myofascial tracks
Middle third of the clavicle, costal cartilage, thoracolumbar fascia, iliac crest	1	
	2	Pectoralis major, latissimus dorsi
Medial line of the humerus	3	
	4	Intermuscular medial septum
Medial epicondyle of the humerus	5	
	6	Group of flexor muscles
	7	Carpal tunnel
Palmar surface of the fingers	8	

3) Deep Back Arm Line (DBAL)

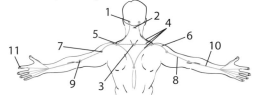

Bony stations		Myofascial tracks
Spinous processes of the lower cervical spine and upper thoracic spine C6–7/T1–4, transverse processes C1–C4	1	
	2	Rhomboid major and minor, levator scapulae
Medial margin of the scapula	3	
	4	Rotator cuff muscles
Humeral head	5	
	6	Triceps brachii
Olecranon of the ulna	7	
	8	Fascia along ulnar periosteum
Ulnar styloid process	9	
	10	Ulnar collateral ligaments
Triquetral bone, hamate bone	11	
	12	Hypothenar muscles
Exterior of the pinkie	13	

4) Superficial Back Arm Line (SBAL)

Bony stations		Myofascial tracks
Occipital margin	1	
Nuchal ligament	2	
Thoracic spinous processes	3	
	4	Trapezius
Spine of the scapula, acromion, lateral third of the clavicle	5	
	6	Deltoid
Deltoid, tubercle of the humerus	7	
	8	Intermuscular lateral septum
Lateral epicondyle of the humerus	9	
	10	Group of extensors
Dorsal surface of the fingers	11	

Posture and Movement

The position of the elbow affects the thoracic spine, while the incorrect positioning of the shoulders could have negative repercussions on the ribs, neck, and breathing.

You will find the FEEL IT and USE IT sections for the Arm Lines in chapter 4, which covers the FEEL IT strategy.

Compensations in the Arm Lines

The following postural patterns are often associated with the Arm Lines:

- Incorrect position of the shoulders
- Carpal tunnel problems
- Shoulder and elbow impingement
- Chronic tension of the shoulder muscles

Arm Lines and Training

AL training should encompass the following exercises:

- Eccentric, isometric, and concentric strength training
- Elastic strength training
- Flexibility
- Three-dimensional movements
- Balance between the four lines

Further considerations for training purposes should also be born in mind in this case. As expected, when moving it is difficult to distinguish between the individual lines. The exercises tend to instruct you to differentiate between the front and back lines so as to correctly balance and distribute the forces, thereby improving movement technique. We shall particularly focus on the opening of the front Arm Lines and on strengthening the back lines, initially performing closed chain exercises before moving on to open chain exercises.

Another interesting aspect is that our postural systems, including our arms, are closely connected to our eyes.

Focus On

To better understand the four lines described above, we shall compare our arms to the wings of a seagull.

Superficial Back Arm Line	
The trapezius and the deltoid form the upper part of the wing, which raise and extend it.	Imagine that you are a seagull; the beat of your wings begins from this line.
Superficial Front Arm Line	
Pectoralis major is the front part of the wing: the driving force of flight, as in geese or ducks.	It is the return phase of the seagull's flapping wings, which controls the speed of flight.
Deep Front Arm Line	
This is the anterior fascia of the wing, which controls the trim.	Imagine that you are the seagull; to dive, you need to slightly rotate your thumb and arm inward.
Deep Back Arm Line	
This is the posterior fascia of a bird's wing, which provides motor control for the ailerons of the flight feathers.	You climb from the dive by externally rotating your arms and pushing your pinkie outward.

2.9. FRONT AND BACK FUNCTIONAL LINE (FL)

The FL is the extension of the Arm Lines through the surface of the trunk to the contralateral extremities. Let me explain: The right arm and shoulder are connected to the left side of the pelvis and left leg, and vice versa. As such, the FL acts as an extension, increasing the lever of the arm or leg.

Why did Thomas Myers choose to call it the "Functional Line"?

The FL is not very active when standing in a relaxed position, but it particularly comes into play when doing sports or other activities that make use of the contralateral force connections of the extremities. Think of a tennis player: The power of a right-handed player that enables him or her to add more pace to the ball comes from the left leg and hip.

FIND IT

1) Front Functional Line

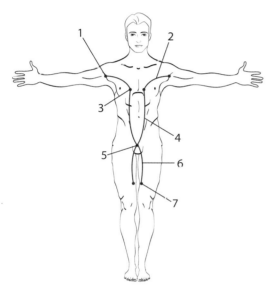

Bony stations		Myofascial tracks
Humeral shaft	1	
	2	Inferior border of pectoralis major
Cartilage of fifth and sixth ribs	3	
	4	Lateral rectus abdominis sheath
Pubic tubercle and symphysis	5	
	6	Adductor longus
Linea aspera of the femur	7	

2) Back Functional Line

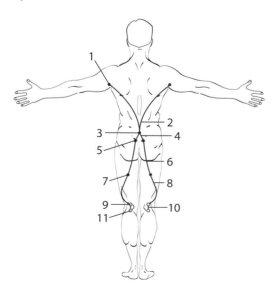

Bony stations		Myofascial tracks
Humeral shaft	1	
	2	Latissimus dorsi
	3	Thoracolumbar fascia
	4	Sacral fascia
Sacrum	5	
	6	Gluteus maximus
Diaphysis of the femur	7	
	8	Vastus lateralis
Patella	9	
	10	Patellar tendon
Tuberosity of the tibia	11	

Posture and Movement

The FLs are not very active when standing upright or in a relaxed position. Instead, they play an important role in stabilizing movements. Many yoga and Pilates poses transmit forces or provide additional limb stability.

The FLs ensure that the movements of the limbs are stronger and more precise, connecting them to the upper or lower extremities through the body. This principle is most evident when jumping, kicking, or playing a sport that requires movement of the arm and contralateral leg. Less evident are the movements of contralateral balance between the shoulders and the sides (hips), step by step, when we walk.

FEEL IT
Explore the FL with the following exercises.

Strengthening of the front and back FL: Push your right shoulder blade toward the left buttock (and vice versa).

Strengthening of the front FL (abdominal X exercise).

USE IT
This is what happens when you throw a ball.

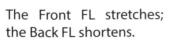

The Front FL stretches; the Back FL shortens.

The Front FL shortens; the Back FL stretches.

Compensations in the Functional Lines
The action of bringing one of your shoulders closer to the opposite side, whether dorsally or frontally, often involves the FLs.

Functional Lines and Training
FL training should encompass the following exercises:
- Eccentric, isometric, and concentric strength training
- Three-dimensional movements
- Different rhythms
- Different directions and angles

I won't spend much time on the Ipsilateral Functional Line, except to say that it can be traced from the most lateral fibers of the latissimus dorsi to the lower outer ribs, moving to the posterior external oblique above the anterior superior iliac spine, toward the sartorius muscle, and ending at the condyle of the tibia inside the knee.

Focus On
With regard to the FL, the following points should be taken into consideration:
- It is an extension of the arms and legs.
- It only plays a minimal role in posture.
- It exerts greater strength.
- It allows three-dimensional movements.

FRᴇE: FASCIAL REAL EMOTION

3.1. THE PILLARS OF THE FRᴇE METHOD

My aim is to make you feel good about your body. With my new FRᴇE method, I have solved the puzzle (at least for now): All the pieces fit together perfectly. The FRᴇE method will revolutionize how you think about training. Everything will be possible, but nothing will be as it was before. You will have new goals to achieve.

The FRᴇE method is built on two pillars to create a unique, simple, fun, and scientific training program that guarantees improved performance. These two pillars are based on science and my personal experience.

The FRᴇE Pillars

Personal experience

- Reebok Master Trainer
- Polar Master Guide
- Be Balanced Master Trainer
- Pilates
- Gyrotonic
- Mézières
- Functional Movement Screen (FMS)
- Certified Functional Strength Coach (CFSC)
- International Kettlebell and Fitness Federation (IKFF)
- Functional Range Conditioning (FRC)
- Anatomy Train I/II

FASCIAL
F = The four-dimensional mesh that encloses us and gives us shape

REAL
R = Real and specific work on the fascial tissue

EMOTION
E = Because fascia training is exciting in all senses of the word

Scientific references

T. W. Myers
R. Schleip
C. Stecco
M. Dalcourt
E. Cobb
A. Spina

Intelligent movement nourishes the mind.
FRᴇE is the missing link between the holistic approach and functional training.

The aim of FReE is to reset the integrity of the myofascial system. To see how the method works, try the following test.

Instant check: Foot massage with ball test

1. Stand up straight with your feet hip-width apart. Turn your head to the left and then to the right. Check how far you can turn your head and note any tension.

2. Place your right forefoot on a sensory ball (or tennis ball), keeping your heel on the floor. Gently push your forefoot down onto the ball for 3 seconds and then relax. Repeat 10 times.

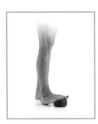

3. Now move the ball to the middle of your foot (your foot should literally wrap around the ball), put all your weight on your forefoot, hold for 3 seconds, relax, and repeat 10 times.

4. Place your heel on the ball. Gently push down through the heel, hold for 3 seconds, relax, and repeat 10 times.

5. Now start rolling your foot over the ball slowly to massage the entire sole of your foot and then increase the speed. Repeat 10 times.

6. Repeat the entire exercise with your left foot.

Has your range of motion (ROM) improved? Do you feel less tension? Great! This is just the beginning. Remember: With FReE the *wow* factor is guaranteed!

3.2. THE FIVE STRATEGIES OF THE FReE METHOD

The FReE method is based on five key strategies that have been extracted and developed from two fundamental pillars: my experience and science.

Why use five different strategies? Because each strategy has a different effect and stimulus on the myofascial tissue. With this method I propose different types of key movements and exercises to comprehensively train your fascia. The training is as varied as the structure of the fascia itself.

As you will have seen in the previous chapters, the structure of the fascia encompasses a variety of structural components that need to be stimulated in different ways (this cannot be missed in training): Use it or lose it!

The movement square, which I introduced in chapter 1 (page 27), represents the complete movement of the fascial tissue and should be born in mind as you read this book. It is provided below once again to remind you of the five key strategies.

FReE incorporates the five movement strategies that the fascia need. As mentioned

The Movement Square

Strategy name	Functions of the fascial tissue
FEEL IT	Perception and activation
MOBILITY	Functional mobility
STRETCH	Fascia stretching and shape
ENERGY	Movement as elastic energy
RELEASE	Fascial release

in chapter 1 (page 27), the fascia communicates, gives shape to the body, moves, and nourishes. As such, FReE uses the five strategies to feel, mobilize, stretch, move, and release (and ultimately improve) the vital functions of the fascia. You will find full details of each strategy at the start of each exercise.

3.3. THE FReE METHOD AND TRAINING

Does fascia impact our daily lives, our fitness, and our sporting activities? The answer is, it certainly does.

Based on the scientific data presented in chapter 1, the better we understand fascia, the more its intrinsic connection with movement becomes clear. Many "muscle" injuries are in fact connective tissue (fascial) injuries, and many muscle injuries occur where the collagen fibers that make up the fascia are depleted.

It is a fact that the muscular system is part of the fascial continuum, and when afflicted by systemic diseases or disorders, its function undergoes a non-physiological change that causes pain. For example, loss of blood flow caused by a wound generates abnormal tension, which can affect the fascial continuum by developing an inflammatory, acute, or chronic environment. Understanding how best to exercise is therefore vital in order to prevent and repair damage and increase elasticity.

The fascia contains 6 to 10 times more sensory nerve endings than muscles. Whether consciously or unconsciously, you will work with your fascia for your entire life whenever you move. What is more, recent studies support the importance of the fascia and of the connective tissue in functional training [17]. Understanding this concept could and should revolutionize our ideas about fitness.

Myofascial training should not replace other forms of training but rather enhance them. As I often say, it was (or still is) the missing link between Pilates and functional training. The chain is now complete, at least for the moment, thanks to the integration of the components that were in large part missing.

FReE is therefore essential training that should complement a strength, cardiovascular, and coordination program. It will complete your comprehensive training program.

This is applicable to both professional and amateur athletes who are looking to maximize their performance, prevent injuries, and improve their post-injury recovery, as well as anyone who wants to feel good in his or her own skin, wake up in the morning feeling great, successfully overcome the challenges of daily life, or simply feel young.

Just one or two intense myofascial training sessions a week are enough for the fascia to reproduce new, fresh, and elastic collagen tissue over the following 72 hours. Alternatively, you can perform individual exercises throughout the day.

3.3.1. Training Plan

A training session can encompass all, some, or individual FReE strategies. It's clear that if the aim is to significantly improve the fascial tissue, all five strategies should be used. The reason for this has already been explained: Myofascial training should be as variable as the fascia itself.

Should training follow a specific order? Yes and no. Let me explain. The five FReE method strategies follow a logical order, but the idea is not to follow a default plan because everyone is very different. It is important to understand why you should make a particular choice, what to do, why you should do it, and how to do it.

It is a good idea to start with the FEEL IT strategy to better feel the movement.

It is important to warm up before loading the fascia in order to make it fluid and improve its functionality and sliding.

Below are some examples of how and where to include fascia training exercises.

1) Warm-up
2) Well-being training
3) Fascia training
4) Athletic preparation
5) Post-training recovery
6) Post-injury recovery

1) Warm Up

Myofascial training can be used in all sessions to warm up, or it can be integrated into an existing warm-up.

Focus:
Optimal preparation for the training to follow. To improve posture and increase flexibility, mobility, and performance, as well as to improve exercise technique.

2) Well-Being Training

These exercises can be used for training designed to improve well-being or can be performed throughout the day. They only take a few minutes.

Plan:
Perform 2 or 3 times a week, selecting exercises from all the strategies or individual exercises that can be repeated every day.

Focus:
To improve daily well-being and posture, as well as to increase flexibility and mobility in daily life.

3) Fascia Training

Intense workout = 30 to 55 minutes, using all the FReE strategies or individual strategies.

Plan:
Perform 2 or 3 times a week, choosing from the different FReE strategies. For comprehensive stimulation, we recommend using all the strategies.

Focus:
To improve general well-being, posture, flexibility, mobility, performance, and exercise technique.

4) Athletic Preparation

Plan:
Integrate FReE into your athletic training, use it to warm up, include some exercises in your weekly training, or even incorporate an entire block of myofascial exercises into your regular training.

Focus:
To enhance performance and minimize injuries.

Patience is called for at the beginning, but after a few months elasticity will increase and the fascial tissue will strengthen, while after one or two years your fascial system will be renewed and rejuvenated.

Feel is everything. With each exercise, ask yourself: How could I perform this movement in a lighter, freer, or more elastic way? Avoid mechanical, robot-like movements. Remember: **Moving is great, but moving and feeling is even better**.

5) Post-Training Recovery

Plan:
Incorporate some RELEASE and STRETCH exercises at the end of an intense training session (e.g., strength training) or after a run.

Focus:
To speed up recovery.

6) Post-Injury Recovery

Having completed the cycle of treatment after an injury, I would recommend adding a few new exercises. It will of course depend on the type of injury sustained. You will find numerous shoulder, knee, and hip exercises in the book.

Plan:
Incorporate some MOBILITY, RELEASE, and STRETCH exercises into your daily life and your training.

Focus:
To help your recovery.

You will find some examples of training programs at the end of the book.

3.4. TEACHING TECHNIQUE

Correct technique provides a foundation upon which to construct your movements. Below are the key teaching principles to ensure a correct technique. By "correct technique", I mean

- correct starting position and
- efficient, safe, and effective movement

Why is it so important to assume and maintain the correct position? There are various reasons:

- To optimize training
- To optimize results
- To maximize improvement
- To improve technique
- To minimize training-related injury

As a trainer, it is very important that I immediately understand how my student compensates. There are thousands of compensations, but it is the little details that ensure correct execution. Pay attention to exercises that appear simple: The devil is in the detail.

By way of example, let's analyze the quadruped position because it presents more points of discussion.

3.4.1. Unbalanced Position

I often see these compensations when people assume the quadruped position as the starting position or during movement. The unbalanced position creates numerous imbalances:

- It loads the joints.
- It does not encourage us to change.
- It requires more effort.

1   Knees and hands not aligned.
2   Knees wider apart than the hips.
3   Hyperextended elbows.
4   Increased lumbar curve.
5   Loss of thoracic curve.
6   Increased cervical curve.

Lack of Alignment Between Lower and Upper Limbs

Limbs that are not aligned do not ensure the optimal distribution of force on the myofascial lines and load the passive structures. The effects of this on the spinal column include the following:

- Increased lumbar curve (lumbar hyperlordosis)
- Increased cervical curve (cervical hyperlordosis)
- Loss of thoracic curve (thoracic kyphosis)

Hyperextended Elbows/Work in Progress

If you extend and lock the elbow joint beyond its physiological limits, communication between the myofascial lines of the arms is blocked, creating an imbalance of the forces. I call this lack of communication in which transmission is blocked "roadblocks" or "work in progress".

The work in progress must first be resolved before communication on the entire line can be established. Problems associated with hyperextended elbows include the following:

- Joint instability increases.
- Joints like the wrist and shoulder are loaded.
- The surrounding ligaments and muscles are weakened.
- If you don't move, your habits won't change.

The joints most affected are the knee, shoulder, elbow, wrist, and fingers.

Work in progress can also be understood to mean a lack of communication on the myofascial lines. For example, think of not being able to activate the latissimus dorsi muscle because communication on this line has been interrupted due to the non-use of certain areas of the body, incorrect posture, or hyperextension of a joint.

Non-Active Shoulder Blades

The shoulder blades are facing the head and are close together. In this position they are unable to transmit force on the myofascial lines, thereby compromising the passive structures like the shoulder joint.

Spinal Column

Because the physiological curves in this position are accentuated, structures like ligaments and intervertebral discs are loaded and the joints are blocked. Performing the exercise in this unbalanced position increases the load on the passive structures.

If the position is balanced, communication between the myofascial lines and appropriate communication on the entire line (fascia and muscle) is guaranteed. At the same time, we will give our body correct balance, equilibrium, and feel inputs. This will help us to change our habits.

3.4.2. Balanced Position

Alignment of Lower and Upper Limbs

- The knees are perpendicular and hip-width apart to ensure optimum distribution on the myofascial lines.
- The hands are shoulder-width apart and perpendicular to the shoulders to ensure optimum distribution of weight across the joints.
- The fingers are spread: Each finger has its own function and communicates with the body.

Outstretched Arms

They help you to communicate and distribute load on the myofascial lines of the arms. Note that I'm talking about straight, not hyperextended, elbows.

Active Shoulder Blades/Active Kite

The active shoulder blades are pushed toward the buttocks (depressed): This is the usual way to explain the position of the shoulder blades. Let's try to take this thought a little further. If we analyze the image of the superficial back line of the arms, we can see that it resembles a kite which begins at the base of the skull (occipital margin), expands to both shoulders, and comes together further down. It consists of four opposing points.

Move the base of the skull (occipital margin) away from the shoulder blades; the collar bones will move away from each other. Push the bottom part of the shoulder blades downward. To create the volume and the expansion, move the thoracic vertebrae away from the sternum; this will create stability in the scapulohumeral joint.

Keep this image in your mind and visualize it whenever I use the term "kite".

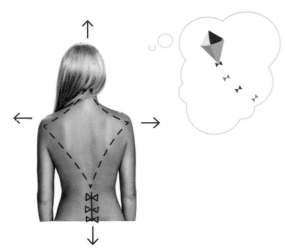

To find the correct position of the humeral head in the glenoid cavity, perform the following movement, which I usually call "screwing in a light bulb".

Raise your right arm to shoulder height. Imagine you are screwing in a light bulb with your right hand by rotating your arm outward. Now try to screw in the light bulb without moving your hand or elbow (which is pointing outward). Movement decreases because only the humeral head rotates in the glenoid cavity. This creates space between the humeral head and the acromion (if they get closer together, tension on the supraspinatus tendon could increase).

Elbows Facing Outward/Rotated Humerus

I have joined these two points because they are intrinsically linked to one another. They help to distribute force between the frontal and posterior myofascial lines of the arms. The position of the elbows affects the shoulder blades and therefore the thoracic spine. Give it a try: Start with your arms by your sides then rotate them outward. You should feel your shoulder blades moving closer together, thoracic kyphosis straightening, and a little tension in the triceps brachii (see figure on the left below).

Change the stimulus: With your arms extended by your sides, move your elbows outward, keeping the collar bones open; you should feel greater activation of the triceps brachii, tension under the lower point of the shoulder blades (activating the serratus anterior and latissimus dorsi), and the shoulder blades moving away from each other while maintaining thoracic kyphosis (see figure on the right below).

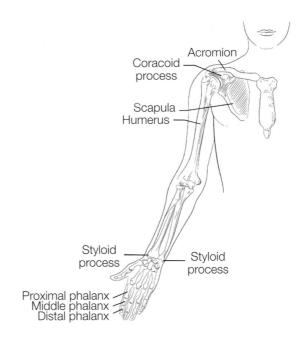

Acromion
Coracoid process
Scapula
Humerus
Styloid process
Styloid process
Proximal phalanx
Middle phalanx
Distal phalanx

Incorrect Position

Elbows facing inward

Elbows facing outward

Correct Position

In light of this information, now adopt a quadruped position. Try to move your elbows toward your knees (external rotation of the arm) just like in the image that shows the incorrect position. You should feel the shoulder blades moving closer to one another, and you should find it difficult to maintain the position of the kite.

Now, try again with your elbows facing outward: Move your shoulder blades toward your buttocks and activate the kite with your elbows facing outward. Turn the humeral head: You should feel tension in the triceps brachii below the lower tip of the shoulder blade, and you may feel the activation of the latissimus dorsi to the iliac crest. This will balance and stabilize your scapulohumeral joint and the upper part of your chest.

The Latissimus Dorsi

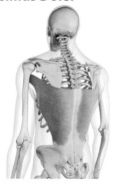

The latissimus dorsi originates from the spinous processes of the last six thoracic vertebrae, the thoracolumbar fascia, the sacral crest, the iliac crest, and the lower lateral ribs; the distal insertion is in the bicipital groove of the humerus.

3.5. TEACHING LANGUAGE

A true coach can help you achieve your set goals with his or her eyes closed. To do so, the coach uses specific language and images to help you get into the correct starting position and to quickly and correctly support your movement. The skill of a coach lies in the ability to adapt the language and choose the right words at the right time. That is why I have developed a specific language to facilitate learning of the FReE training method.

3.5.1. Contact Points, Points of Support, and Conductors

What do we mean by the term "contact points"?

Let's take the quadruped position as an example. In this case, resting your hands on the floor marks the start of communication. Contact between your hands and the ground stimulates the peripheral receptors, transforming the mechanical stimuli (from the surface of your hand) into nerve impulses. These are then transmitted by the sensory nerve fibers to the upper nerve centers, where they are decoded. Simply put, an impulse is sent to the nerve centers. It is therefore clear that contact with the floor stimulates, communicates, and activates the receptors of the hands and the myofascial lines of the arms.

The contact points help us to feel the myofascial lines; to improve the distribution of force on the lines involved; and, at the same time, to balance the lines themselves. All this helps to

- improve posture,
- increase the transmission of force,
- reduce loads on individual joints, and
- reduce exertion.

But be careful not to confuse a contact point with any point of support with the ground. There is a substantial difference that I am going to explain.

Contact Point

In the quadruped position there are two contact points: the hands, because they help you to communicate with the myofascial lines of the arms, and the toes, which help you to communicate with the superficial back line (SBL).

Points of Support
In this example, the knees are a point of support, but not a contact point, because the knee is in the middle of the myofascial line (in this case the superficial front line) and is not able to communicate.

Movement Conductor (Initialization)
"Conductor" refers to the starting point of the movement, and I can assure you that the starting point significantly affects the quality of the action.

Imagine that you are walking down the street when you suddenly spot a famous person. What do you do? Wouldn't you turn your head and look to make sure that it is really him or her? OK, now recreate this exact rotation to your right and make a note of how far you can turn (without too much effort).

Now try it again, but this time start the movement with your right ear lobe: This is the conductor of the rotation. Hasn't working with the conductor improved the rotation? This is precisely the difference: quality, excellence, and the key to intelligent movement.

3.5.2. Fixed, Mobile, and Opposing Points

Fixed, mobile, and opposing points are part of my imaginative language to achieve a specific objective with ease. If I focus on bringing two points in opposition, I don't have to think about moving bone segments or working with individual muscles but rather changing the tension between these two points. Tension follows the shape.

Let's look at a few examples by analyzing the quadruped position.

Fixed Point and Mobile Point
The fixed point is the pubis, while the mobile point is the sternum (fifth rib). In this way, we are transmitting a tension input to the SFL.

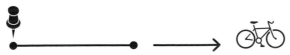

Opposing Points: End to End
The base of the skull (occipital margin) and the coccyx move away from each other. This transmits a stimulus to the deep front line (DFL), thereby lengthening the spinal column. (Note: I am not talking about straightening the physiological curves but rather changing the tension.)

Two Fixed Points and One Mobile Point
The two fixed points are the hands resting on the ground, while the mobile point is the thoracic spine, which moves away from the hands, creating volume.

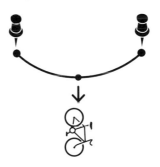

Two Mobile Points and One Fixed Point
This is simply the opposite of what I have just explained.

Many other combinations are of course possible (two opposing points, etc.), but I shall stop here because I don't want to overcomplicate things. I have chosen these three variations because these are the ones I use the most, and because they have an instant impact on achieving the set goal.

3.5.3. Other Terminology

Focus On

You will find a "Focus On" section with every exercise in this book. I use it as a quick summary to remind you of the most important points to focus on as you perform the exercise. I could undoubtedly add some more, but I have chosen to only highlight the most important ones: common errors that I regularly see when working with groups or individuals, who often find the same compensations.

Conscious Distribution of Force

This expression refers to the ability to direct the distribution of load, tension, force, and information to the applicable myofascial lines. This correct distribution of force balances the myofascial lines between themselves and helps you to use the muscles that really should be working.

Screwing In a Light Bulb

The movement that I describe as "screwing in a light bulb" is explained in detail on page 65.

Pre-Tension

Pre-tension is an opposing movement. This is generally true for all fascial structures surrounding the muscles (epimysium and perimysium that wrap around the muscles and tendons), but it is particularly applicable to tendons.

Through this opposing movement, tension on the tendons and ligaments increases and energy is stored. Imagine stretching a rubber band. When you let go of the elastic band, the built-up energy is explosively released.

3.6. THE FOOT IN DETAIL

Quick question: How often do you stop to look at your feet? We generally only look at our feet when they hurt: calluses, hallux valgus, plantar fasciitis, ingrown toenails, and so on. What you may not know is that your feet are like an open

book, and they speak to you and communicate with you when there is something wrong. Why ignore their messages?

Let me give you an example: If the brake warning light in your car illuminates while you are driving, indicating that something is not right, what do you do? Do you ignore it? Most people worry about it and take their car to the garage. The same thing happens with our body; it sends us messages that we often ignore or do not know how to interpret.

For this and many other reasons that we will soon see, in this book I have included numerous exercises to keep your feet beautiful, strong, and healthy.

We will now look at the foot in detail. It comprises

- 3 arches,
- 26 bones (28 if we count the sesamoids; one-quarter of all your bones are in your feet),
- 107 ligaments,
- 33 joints, and
- 33 muscles.

The structure of the foot is designed to support the weight of your body, and it is the only contact point with the ground when we are standing up and on the move. If you counted all the steps that you walk in one day and multiply this by months, years, or a lifetime, you would realize just how hard your feet have to work and how much weight they have to bear.

The foot has a number of functions, all associated with walking, jumping, and running.

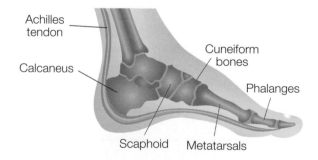

Achilles tendon
Calcaneus
Cuneiform bones
Phalanges
Scaphoid   Metatarsals

- Balance
- Propulsion
- Flexibility

From a functional perspective, the architecture of the foot is divided into three distinct sections:

- **Forefoot**, formed of the toes (phalanges) and the metatarsals.
- **Midfoot**, formed of the bones found in the middle of the foot (comprising the cuboid bone, the three cuneiform bones, and the navicular bone).
- **Hindfoot**, the rear of the foot supported by the calcaneus (heel bone, comprising the talus bone and the calcaneus).

The hindfoot, which comprises the large calcaneus bone, a large bony growth, impacts the ground and absorbs the force from the trunk and the lower limbs. The midfoot has a cushioning action and promotes lymphatic and venous return through the dynamic compression of the plantar vascular structures. The primary function of the forefoot is propulsion. The bones of the foot are held in position, moved, and supported by a fascial network, muscles, tendons, and ligaments.

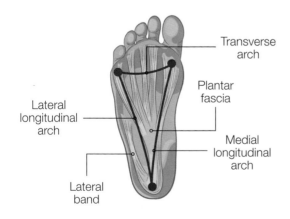

We will now look at the correct posture of the foot and the distribution of weight between these three points.

The foot has three points of contact with the ground:

- **Anteromedial contact point**, which is the head of the first metatarsal
- **Anterolateral contact point**, which is the head of the fifth metatarsal
- **Posterior contact point**, which is on the posterior tuberosity of the calcaneus

When standing, 60 percent of your body weight should be borne by the forefoot and the remaining 40 percent by the heel.

The foot compresses the vascular structures to promote lymphatic and venous return; this is because the lower limbs do not benefit from the direct thrust of a pulsating pump like the heart. There are two particularly important structures inside the foot:

1) Lejars' plantar venous pump: This is a network of capillaries that promotes lymphatic and venous return upward when compressed during walking or running.
2) Plantar vault: This is a structure composed of the following three arches:

   - Medial longitudinal arch: From the calcaneus to the head of the first metatarsal.
   - Lateral longitudinal arch: From the calcaneus to the head of the fifth metatarsal.
   - Transverse arch: From the first to the fifth metatarsal head.

This architectural structure has the task of adapting to the harshness of the terrain and its inclination. In addition, with every step and major movement, the compression of the veins of the plantar vault acts as a peripheral heart, which represents the most important vascular function of the foot. This is why FReE should be done barefoot as we look to find balance in our feet.

The veins of the foot respond to the contraction and relaxation of the muscles of the plantar fascia. This is why you are advised to move your feet (raise and lower your toes or heels) when on a long-haul flight and have to sit for hours. These exercises are also called the muscle pump and prevent dangerous thrombosis (blood clots).

Venous return may be facilitated by graduated compression stockings, which promote venous emptying by acting as an additional external pump. Many athletes use them to enhance their performance. However, I am not asking you to wear compression stockings but to train your fascial structures. In daily life, well-trained and toned fascia performs the same task.

The soles of our feet require a lot of care and attention. They are often the source of problems and tension that are then transmitted upward to the rest of the SBL, as you can see for yourself with the test on page 60.

Problems with the soles of the feet, such as tension and lack of mobility and stability of the joints (often associated with tension in the femoral muscles, lumbar hyperlordosis, or persistent hyperextension of the neck), create problems in the joints directly above, such as the knee and hip.

The plantar aponeurosis (plantar fascia) is like a bouncy trampoline. It forms an elastic arch between the three contact points: the head of the first metatarsal, the head of the fifth metatarsal, and the middle of the heel (calcaneus).

Its tone and elasticity adapts to every situation. Let me explain. If you squeeze your feet into shoes that are too small or tight or into stiletto heels (for a long time), their functionality will be limited. The myofascial tissue adapts and responds with

- plantar tension,
- minimal joint mobility and instability, and
- tension in the calf muscle.

Our entire body adapts to the demands we place on it. It is not without reason that they say that posture is the window to the soul.

Incorrect foot posture leads to a series of unwanted consequences, including the following:

- Incorrect spinal posture
- Ineffectiveness of the lymphatic–venous pump, leading to cellulite and circulation problems (telangiectasia, varicose veins)
- Back pain
- Weak pelvic floor
- Plantar fasciitis
- Hallux valgus
- Flat feet
- Arthrosis
- Muscle imbalance
- Soreness

Remember that many of these negative consequences are caused by wearing heels that are too high (2 inches or higher) or shoes that are too tight.

3.6.1. Heel Spur (Calcaneal Spur)

A heel spur is a common affliction among athletes whose sport of choice involves jumping or running long distances. It manifests as crippling, sharp pain. It is a bone deformity on the underside of the heel; the tip of this bony outgrowth faces the toes and is generally found in the inferomedial part of the calcaneus. It is caused by inflammation where the plantar fascia is inserted on the heel, resulting in the deposit of calcium salts. Over time, this build-up in the heel develops into a heel spur.

Causes can vary, from incorrect foot posture to continuous and repeated tension on the plantar fascia, which constantly pulls on the insertion of the heel bone to the plantar fascia.

This fascia is not really attached to the calcaneus but binds to its surrounding periosteum, like a mesh sock that wraps around the bone. When continuous tension is exerted on this mesh sock, it may come away from the bone, creating a space between the bone and the periosteum (sock). Osteoblasts (cells that make bone) continuously clean and reconstruct the outer surface of the bone. If continuous tension is applied to the plantar fascia, the fascia may come away from the bone. The osteoblasts fill this space, creating a bone spur. Remember that structures are shaped according to their use. The spur does not cause discomfort per se unless it touches the sensory nerve.

To summarize:

- Work on the plantar fascia (with a tennis or a massage ball) and calf muscle (eliminating tension and stretching) with movements found throughout chapters 4 through 8.
- Use the healthy feet programs in section 9.1.
- If it does not improve, consult a podiatrist, osteopath, or physiotherapist and include specific exercises for the plantar fascia.

3.6.2. Hallux Valgus

This condition often arises when the myofascial lines of the feet are unbalanced or due to shoes that are too tight or heels that are too high. In the initial stage, working to rebalance the foot can lead to an overall improvement. If the problem persists over a long period of time, improvement will be more difficult to achieve. Hallux valgus may be the most common condition to afflict women. Although its cause may be weight or age related, it can also be genetic and is often found to be hereditary. It is a deformity by which the base of the first metatarsal points laterally to the head of the first phalanx. This creates the classic "onion", the stage and severity of which can vary depending on the deformity, redness, and pain.

Remember: Function follows the shape. Please see the healthy feet programs in section 9.1.

FEEL IT: PERCEPTION AND ACTIVATION

FEEL IT is a strategy that uses a range of exercises to increase the number of files in our computer.

More motion, more software.

4.1. FEEL IT STRATEGY

Let me introduce you to the different nuances of the FEEL IT strategy (which involves listening and feeling), from simple to more complex exercises. Experiment with your body and explore its potential in a different way; this will help you to consciously feel what is happening. Only then will this awareness be deeply rooted in your brain, and only then will your training truly change.

Start with this strategy because moving is great, but moving and feeling is even better. The time has come to start to listen to and feel

73

your body, which requires constant stimuli to maintain and grow intelligent movement, as we saw in chapter 1.

In this chapter I explain in detail why and how I came to include the FEEL IT strategy in my training sessions.

Let's consider precisely what is meant by proprioception and perceptions. I prefer the definition proposed by the neurophysiologist Charles Scott Sherrington (1857 to 1952) because it is the definition that I most often come across and which I have adopted for my courses. Sherrington first coined the term "proprioception" in 1906.

Definition of Proprioception

The word derives from "propius" (oneself) and "(re)ceptus" (to take or grasp), or "the ability of our body to take or grasp itself". Sherrington then went on to define the "kinesthetic sense" as kin(ema) (movement) and cistes(is) (feeling or sense). The proprioceptive sense is linked to sensory inputs that originate in fascial tissue in the tendons, joints, aponeuroses, ligaments, muscles, etc. (see chapter 1).

Sensations and Perceptions

Sensations are electrochemical impulses that originate from sensory detection, which are processed in order to be perceived.

Perceptual Learning

"Perceptual learning" is defined as the acquiring of new behavioral models, the establishment of new response configurations adapted to environmental (external and internal) and individual requirements.

Simply put, the more we feed our brain, the better it will be able to process inputs (afferent stimuli) and respond in terms of outputs (efferent stimuli).

Applying the FEEL IT strategy will help you to consciously feel what is happening in your body. As a result, this learning will be deeply rooted in your brain and will then be able to

trigger changes, offering you the following benefits:

- Awareness of movement
- Safety in movement
- Improved functionality of everyday movements
- Improved athletic performance
- Greater muscle flexibility
- Stimulation of the nervous system (by communicating with the brain, it nurtures the mind, creating new files)

In a nutshell, perception is essential for complete mental and physical well-being, as well as for athletic performance; you need constant stimuli. Unfortunately, this component is very often neglected or put to one side because it is not easy to listen to oneself.

Using the contact points (see section 3.5.1) will help you to feel the myofascial lines. Conscious distribution allows us to

- direct the distribution of load, tension, force, and information to the applicable myofascial lines, and
- have the perception to connect or disconnect the myofascial lines and therefore to consciously influence them.

4.2. ON/OFF EXERCISES

The ON/OFF exercises re-educate the muscles so that we can both activate and deactivate them.

For various reasons (postural, emotional, and others), a muscle very often does not relax because it is too stressed, or because it has been contracted for a long time and has been in a state of sympathetic nerve activation for too long. In order to relax the muscle and bring it under the influence of the parasympathetic nervous system, it first has to be reactivated.

Muscle tone means reflex and constant muscle activity that maintains the postural structure of the body by opposing the force of gravity. It is also defined as the minimum tension required by a particular group of muscles to complete an action.

The neuromuscular reactivation exercises stimulate the conscious and unconscious conversation between the central nervous system on the one side and the muscles and fascia on the other.

Objective

The purpose of these exercises is to activate, reactivate, and reset the neuromuscular mechanisms that control various components of the structure of our body, particularly when we are standing upright, stationary, or in motion.

Example

Let's work on the iliopsoas. Flex your hip to 90 degrees and isometrically contract your quadriceps muscles for about 3 seconds. Then relax the muscle by straightening your leg. Repeat the exercise 3 times.

Now work the same muscle with your leg straight (thigh extended). Isometrically contract the muscle for 3 seconds. Repeat the exercise 3 times.

This will work on the brain's inputs to improve its outputs (its ability to respond).

The contribution of Henri Laborit (1979) is important to this concept. His theory on the mechanism of inhibition of action asserts that everything that cannot be expressed in the human being can be transformed into a pervasive and chronic block, with negative repercussions on self-control and psychophysical health. Peter Levine refers to this block when he describes an impala being chased and caught by a cheetah [40]. The Impala goes limp and falls to the ground as soon as the cheetah pounces, appearing to be dead, but what is actually happening to the impala's body is similar to what happens when you put your foot on the gas pedal and brake of your car at the same time. The antagonism between the internal workings of the nervous system (engine) and the external immobility of the body (brake) leads to intense agitation within the body, similar to a storm. If the impala escapes, it will discharge all its mobilized energy by shaking and quivering wildly to regulate its nervous system again. However, humans often omit this step, and the trauma remains trapped in the body.

Feel, change, and integrate: You should keep this common thread in mind when you exercise. Small movements and subtle changes of position or direction are required.

4.3. EXERCISES FOR THE FEEL IT STRATEGY

In the following sections you will find exercises for the FEEL IT strategy subdivided as follows:

- Exercises against the wall (exploring the lines and their activation)
- Bodyweight exercises
- Exercises with a foam roller and balls

4.3.1. FEEL IT Exercises Against the Wall

For this series of exercises, the wall, which is both loved and hated in equal measure, will be your training partner.

Standing Square Against the Wall (SBL)

Starting Position

Standing up straight, place your hands on the wall at chest height and slightly more than shoulder-width apart and lean toward the wall.

Movement

Take two steps backward and lower your chest until it is horizontal to the floor so that your body makes the shape of a square (with the wall and floor). Keep your head between your arms, your back in a neutral position, and your shoulder blades pushed toward your buttocks.

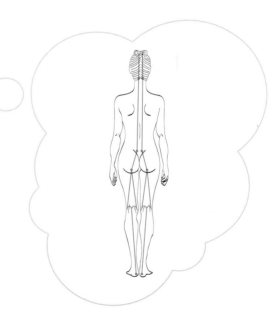

FIND IT

Mentally trace the path of the SBL, connecting one point to the next.

FEEL IT

- Try to feel the individual parts of the SBL. Use the points to help: Fixed point: heels; mobile point: hips.
- Raise the hips, which become a mobile point, toward the ceiling; tension in the back of the thigh (hamstring) will increase. Now bend your knees slightly and the tension will decrease or even disappear.
- With your legs straight, the SBL acts as a continuous myofascial line.

USE IT

Hold the square position to the wall for 30 to 45 seconds and use the SBL in a static stretch.

Focus On

- Active kite.
- Elbows facing outward.
- Outstretched arms.
- Maintaining your back and pelvis in neutral position.
- Palm of your hand completely flat against the wall.

Work On

Activation and perception of the SBL.

Discussion

This exercise can be used to feel the lower part of the SBL or to activate the SBL. Adopt the position, hold it for 3 seconds, and return to the starting position. Repeat 4 to 10 times.

Change Stimulus

Roll down the wall. To feel the top of the SBL, lean your pelvis, thoracic spine, and head against the wall. Breathe out and start to gently push your cervical vertebrae against the wall. Stretch your neck and lower your head in front of your chest toward the floor, removing one vertebra at a time from the wall but keeping your sacrum against the wall as much as possible and your coccyx toward your heels. The tension (feeling of stretching) in your back will increase as you go down; this is the top of the SBL.

If you reach the point where your vertebrae come off the wall in groups rather than individually, roll back up to the point where the vertebrae were still rolling off the wall one at a time. Pause here, take three breaths, and try to roll back down a bit further.

Allow time for the fascial network to adapt. Breathe in as you slowly return to an upright position. Repeat the exercise multiple times, and you will be surprised to see how fast your flexibility improves. The fascia adapts to the demands we put on it.

Reach Back to the Wall (SFL)

Starting Position

Stand up straight with your back facing the wall, about 12 inches away. Your legs are shoulder-width apart, arms by your sides.

Movement

Slowly raise your right arm above your head and place it on the wall.

FIND IT

Mentally trace the path of the SFL, connecting one point to the next.

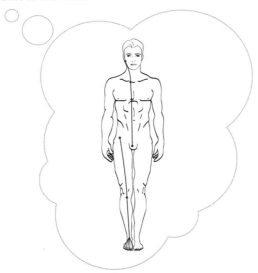

FEEL IT

Push your heels against the floor and your right hand against the wall, push your pelvis forward, move your fifth rib away from your pubis, expand your chest toward the ceiling, and move your head away from your shoulders. Feel the SFL stretching, particularly the right side.

If you feel the lumbar spine compressing, take one step backward and stretch up before you reach the arm back, trying to arch like a rainbow. Hold for a few breaths, return to the starting position, and repeat with the left arm.

USE IT

Hold the position for 30 to 45 seconds and use the SFL in a static stretch.

Focus On

Maintain the expansion in your whole body.

Work On

Feel the static stretch of the SFL and Arm Lines.

Discussion
This exercise is excellent preparation for the full bridge.

Reduce the Level of Difficulty

- Move closer to the wall, thereby reducing the ROM.
- Lean your back against the wall instead of your hand.

Increase the Level of Difficulty

Raise both arms.

Strategy

FEEL IT – Perception – Exploration and Activation – Arm Lines

Starting Position

Standing up straight, bring your toes close to the wall and raise your outstretched arms to shoulder height, with your elbows facing backward. Place the palms of your hands on the wall, maintaining a gap between the wall and your chest. Hold your back and pelvis in neutral position.

FIND IT

Mentally trace the path of the SFAL, connecting one point to the next.

FEEL IT

Press your fingers and palms gently against the wall to create pre-tension on the Superficial Front Arm Line. This will help you to feel its path. If you concentrate, you should feel the activation of the pectoralis major and latissimus dorsi.

USE IT

Hold the position for 30 seconds to use the exercise as a perception tool.

Focus On

- Active kite.
- Elbows facing backward.
- Outstretched arms.
- Shoulder blades open.
- Back in a neutral position.

Work On

Activation and perception of the Arm Lines.

Discussion

This and the exercises that follow are very simple and can be performed anywhere.

You will only feel the activation of the myofascial line paths if you adopt the correct position: shoulder blades active and correct position of the elbows. Why are the elbows so important? Because they affect the position of the shoulder blades. The shoulders have an impact on the ribs.

I bet that now that you have seen the exercise, you can't see the point. I understand, but you are wrong. These activations are vitally important for the bodyweight exercises that you are going to do, such as movements in the quadruped or plank position, or exercises with barbells or kettlebells, because the contact points activate the myofascial lines to ensure the correct distribution of force, balancing one line against another and helping you to use the muscles that you really should be working.

Use these types of movement in your warm-up.

Test

Adopt all the positions of your arms on the wall (described in variations 1, 2, and 3) but only with your right arm. Then step away from the wall and raise both arms above your head. Do you notice a difference? Normally the arm that has been activated is lighter, more active, and more reactive.

Change Stimulus

Use this exercise as ON/OFF activation. Starting from the same position, the only thing that changes is the activation. Hold the position for 3 seconds to activate the pectoralis major and latissimus dorsi, relax for a few seconds, and repeat 5 to 10 times. Continue with all the positions of the arms (described in variations 1, 2, and 3). The aim is to activate the myofascial line and relax ON/OFF to rebalance the function of the muscles.

VARIATION 1: Activation of the Deep Back Arm Line

FIND IT

Mentally trace the path of the DBAL, connecting one point to the next.

FEEL IT

Rotate your outstretched arms outward, keeping your elbows facing backward, and place the external edges of your hands (pinkies) against the wall. Press gently against the wall, and you should feel the activation of the triceps brachii and rhomboid muscles on the Deep Back Arm Line.

USE IT

Use this exercise as ON/OFF activation.

VARIATION 2: Activation of the Deep Front Arm Line

FIND IT

Mentally trace the path of the DFAL, connecting one point to the next.

FEEL IT

Internally rotate your outstretched arms, adopt the kite position by keeping your elbows facing backward, and place your thumbs against the wall. Push gently against the wall, and you should feel the activation of the biceps and the pectoralis minor on the Deep Front Arm Line.

USE IT

Use this exercise as ON/OFF activation.

VARIATION 3: Activation of the Superficial Back Arm Line

FIND IT

Mentally trace the path of the SBAL, connecting one point to the next.

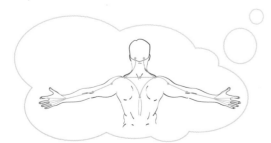

USE IT

Use this exercise as ON/OFF activation.

FEEL IT

Standing up straight, lean your sacrum, back, and head against the wall. Adopt the kite position, keeping your spine and pelvis in a neutral position. Raise your arms to shoulder level, with the back of your hands against the wall. Your elbows must be facing backward toward the wall.

Push your fingers and the back of your hands gently against the wall to activate the finger extensor muscles, the deltoid, and the trapezius muscle on the Superficial Back Arm Line.

Strategy

FEEL IT – Perception – Exploration and Activation of the Arm Lines

Starting Position

Standing up straight and with your back leaning against the wall, ensure that your pelvis and spine are in a neutral position (the sacrum, thoracic spine, and head should be in contact with the wall). Your arms are straight by your sides, with the palms of your hands against the wall, elbows facing outward. Hold the correct position and breathe for 30 seconds, relax for a few seconds, and repeat 4 more times.

Focus On

- Elbows facing outward.
- Active kite.
- Thoracic spine (T12) and head against the wall.
- Lumbar spine and cervical spine away from the wall.
- Shoulder blades open.

Work On

- Perception of the Arm Lines.
- Fascia stretching of the arms and chest: SBL/DFL/AL.

Discussion

Gently press the outer edge of your hand and your pinkie against the wall without lifting your thumb off the wall. You should feel the triceps brachii activating. Activate the shoulder blade (by pushing it downward slightly) and feel the latissimus dorsi muscle to the iliac crest. This discussion will continue in the next movement.

Change Stimulus

Back to the wall with arms raised: Once in the correct position, activate the shoulder blades and the latissimus dorsi. Slowly raise your arms, keeping them shoulder-width apart, with your thoracic spine (T12) against the wall. Raise your arms as far as they will go while ensuring that the rest of your body does not deviate from the starting position; otherwise, you will not achieve the desired stretch, and it will be a waste of time.

Focus On

- Elbows facing outward.
- Active kite.
- Screw in the light bulb (external rotation of the humeral head).
- Fixed point: thoracic spine (T12); mobile point: hands

Work On

Dynamic stretching of the Arm Lines and chest.

Discussion

This exercise, or homework, if you will, is a must for me since it only requires a wall and two minutes. There is efficiency in simplicity because it allows you to feel exactly where the retractions or, as I like to say, "works in progress", occur.

The tendency will be to raise your arms too far, dragging the rib cage with them, closing the shoulder blades, and displacing the humeral head into an unfavorable position.

Here is a quick list of some common errors and compensations that come to light:

- It is impossible to keep your elbows facing outward without changing the shoulder blades.
- It is impossible to keep your wrist against the wall.
- Your head does not touch the wall.
- Your head touches the wall, but the cervical curve increases.
- The collar bones close forward.
- In trying to hold T12 against the wall, retroversion of the pelvis tends to occur.

So the question is this: Why not spend a bit of time doing such a simple exercise to improve everything else (posture, breathing, and your entire workout)?

Stork Against the Wall (LL)

Starting Position

Stand close to the wall, feet hip-width apart, arms straight and down by your side.

FIND IT

Mentally trace the path of the LL, connecting one point to the next.

FEEL IT

To feel and activate the Lateral Line, gently press the outside of your foot, knee, and hip against the wall.

USE IT

• Hold the position for 30 seconds and then change sides.
• Use this exercise for ON/OFF activation.

Movement

Raise your right arm to shoulder height and ensure it is touching the wall. Raise your right knee up to the height of your pelvis. Make sure that the outer edge of your foot and hip are touching the wall. Bring your knee as close to the wall as possible.

Focus On

Keeping your foot, knee, and hip against the wall.

Work On

Perception and activation of the LL.

Discussion

This is a fantastic, instant, and simple exercise. It allows me to use the acronym KISS – *Keep It Simple Stupid*.

Try the test below; the *wow* factor is guaranteed. There really is nothing else to say.

Test

Before changing sides, listen to your body. Is your right side different from your left?

Move away from the wall, raise your right leg (Stork pose), and then repeat with the left leg. Has anything changed? Your leg should be lighter, more stable, and more active.

Stand on your right leg and squat then change to your left leg. What a difference!

Lean to your left side and then to your right side (see the first figure at the top right of page 45), and analyze the movements. Are they the same?

Do you now see how useful this exercise is?

Strategy

FEEL IT – Activation of the Gluteus Maximus (LL)

Starting Position

Stand with your right side toward the wall, 4 inches (10 cm) away from it. Take half a step backward with your left foot, keeping your forefoot on the ground and your heel raised. Keep your arms straight and down by your side.

Movement

To start, push the big toe of your left foot into the ground. Push the entire sole of your right foot into the ground, as if trying to slide it forward. Once you have created this tension in your legs, gently flex your hip by moving your pelvis backward and toward the wall, with your back extended. Move your left arm diagonally away from your chest, thereby activating the back FL (from the latissimus dorsi on the left side of your body to the gluteus maximus on the right). Hold for 3 seconds before returning to an upright position. Repeat 10 times and then change sides.

Discussion

I first saw this exercise performed by the legendary Michol Dalcourt (creator of ViPR), an internationally renowned trainer who has the ability to explain the most complicated concepts with amazing simplicity that completely reflects him as a person.

I promise you that this exercise is a great way to reset the gluteus maximus ON/OFF, and you will feel its effects.

Focus On

- Fixed point: big toe and foot; mobile point: pelvis and arm.
- Active kite.

Work On

ON/OFF activation of the gluteus.

Change Stimulus

- If your right knee tends to twist, move your left arm toward your right knee to feel the magic. This will pull your right knee slightly outward, giving you the *wow* factor. At the same time, you will feel the tension in your right buttock increase.

- If you don't have a wall, you can do this exercise without one.

Increase the Level of Difficulty

Perform the same exercise using a medicine ball or a ViPR to increase the load. Hold the position for 3 seconds and then return to the starting position to relax. This will undoubtedly increase the involvement of the abdominals, so why not give your core a complete workout?

Strategy

FEEL IT – Exploration – Perception and Activation of the LL

Starting Position

Stand sideways to the wall, about 5 feet (1 1/2 meters) away, feet hip-width apart, arms straight and down by your side.

Movement

Take one step toward the wall, bend your hip (moving it slightly behind your buttocks) and your knee, and place your hands on the wall in line with your chest.

FIND IT

Mentally trace the path of the left LL, connecting one point to the next (see the pink dots in the figure above).

FEEL IT

Explore the entire LL by stretching. Push the outside of your left foot in one direction and the pinkie of your left hand in the other.

Explore and strengthen the lateral line. Push your right hand gently against the wall and move your torso slightly away (creating an expansion on the left side of your torso).

USE IT

Hold the position for 30 to 45 seconds and then change sides.

Focus On

- Opposing points: left foot, left hand.
- Active kite.

Work On

- Perception and activation.
- Muscle stretching and strengthening.

Increase the Level of Difficulty

As you move away from the wall, place your legs slightly behind your torso, increasing the rotation.

Change Stimulus

- Use this exercise as ON/OFF activation. Adopt the position, hold it for 3 seconds, and then return to the starting position, activating the LL each time by stretching and strengthening.
- Use the exercise in the ELASTIC ENERGY strategy. From the starting position, adopt a side step position while leaning your arms against the wall, then return to the starting position with an elastic and explosive movement.
- As an alternative to the previous point, raise your right leg to assume the Stork pose before returning to the starting position.
- Perform the variation described in the point above then, from the Stork pose take a step forward with your right foot, moving into a dynamic front lunge position. Immediately return to the Stork pose to then repeat it from the beginning. Exploit pre-tension and counter-rotation in this movement. Repeat 5 to 10 times and then change sides.

4.3.2. FEEL IT Body Weight Exercises

Forced Breathing

Strategy
FEEL IT – Activation – Breathing

Focus On

- Forced expiration.
- Active kite.

Work On

- Activating the sympathetic autonomic nervous system.
- Movement preparation.

Starting Position and Movement

Lie down on your front, place your hands under your head, and activate the kite. Breathe in through your nose and breathe out forcefully through your mouth, making a loud "CHHH" sound. Repeat 10 times.

Discussion

This exercise is not suitable for people with high blood pressure.

We focus on forced exhalation to activate the sympathetic autonomic nervous system, which helps prepare us for training. Try using it as a warm-up exercise before exercise. You should feel ready, fully charged, active, and full of energy.

Breathalyzer Breathing

Strategy
FEEL IT – Activation – Breathing

Starting Position and Movement

Stand up straight. Breathe in. Then breathe out three times without inhaling in between. On the third exhale, expel all remaining air from you lungs. Try to make the sound "HHHH" as you exhale.

Dog Breathing (Panting)

Strategy
FEEL IT – Activation – Breathing

Starting Position and Movement

Stand up straight as in the previous exercise. Rapidly breathe in and out through your mouth for 15 to 30 seconds. You may have noticed that dogs use the same fast breathing when they are stressed.

Discussion

This type of breathing is stressful for the body; the pupils dilate, heart rate increases, the gastrointestinal system is impeded, muscle strength increases, and the body releases catecholamines, particularly adrenaline. In other words, our body is on alert, ready for attack.

It is not a type of breathing to be activated randomly, but if used at the right time, it can become positive stress for our body. However, breathing like this the whole day would result in negative stress with numerous repercussions for the body, including posture.

Breathing shall be rebalanced in the release phase by activating the parasympathetic nervous system. During the day, the sympathetic and parasympathetic nervous systems should be balanced, and they should activate in ON/OFF mode at the right time. This **constant balance** between sympathetic and parasympathetic components allows us to rapidly adapt to any and all changes that we may face in daily life, and it is usually a determining factor for good health.

Starting Position

Stand up straight with your left arm raised in front of your chest at shoulder height and your index finger pointing upward. Look at your finger.

Movement of the Arm

Fixed point: head; mobile point: finger. Follow your finger with your eyes as it moves to the left, keeping your head still. Repeat 5 to 10 times and then change sides.

Movement of the Head

Fixed point: finger; mobile point: head. Turn your head to the left, keeping your eyes fixed on your finger. Repeat 5 to 10 times and then change sides.

Focus On

- Keeping the fixed point completely still.
- Turning your head or move your arm as far as you can without losing sight of your finger.
- Active kite.
- Eye movement (neurological and motor function).

Work On

- Stimulating the nervous system.
- Stimulating coordination.
- Stimulating proprioception.
- Reducing tension in the neck.
- Improving vision.

Discussion

Exercises like this should always be part of your training.

"The **eye** is a tremendous instrument that transmits to the brain everything it needs to understand an individual's surroundings. **Sight** is simply a response to the light; it is the objectively measurable reflex action of the eye turning toward the light. **Vision**, on the other hand, is the process of interpreting what the eye sees, a process that extracts meaning from what we see, the ability to understand and integrate what we have seen, bringing into play all the senses: touch, hearing, taste, and smell. All this is mediated by an individual's emotions, surroundings, and imprinting received. Vision forms the cornerstone of life because around 80 percent of all the information that is useful for an individual's survival and development is received by sight, while the other senses account for the remaining 20 percent. Sight is therefore the most important of the five senses" [10].

Feed the mind and the output changes.

Don't believe me? Let's do a before and after test: Turn your head to the left and then to the right, making a mental note of how far you can turn and the tension that you feel. Now do the Movement of the Arm exercise and repeat the test. Has anything changed?

Change Stimulus

- As you turn your head, first close one eye and then the other.
- As you turn to the left, close your right eye and push your tongue against the right side of your mouth.

Starting Position

Stand up straight with your left arm raised in front of your chest at shoulder height and your index finger pointing upward. Look at your finger.

Movement: Side Lunge
With Movement of the Head

Fixed point: finger; mobile point: head.
Perform a side lunge while keeping your eyes on your still finger. Let's stop a moment to talk about the side lunge technique: Take a step to your right and bend your hip, moving your buttocks backward slightly and, as a result, slightly bending your right knee. Keep your left leg straight and your back in a neutral position and as straight as possible. Return to the starting position. Start at a close distance before gradually increasing.

Movement: Side Lunge
With Movement of the Eyes

Fixed point: finger and head; mobile point: eyes. Perform the side lunge while keeping your eyes on your finger, without moving your finger or your head.

Movement: Side Lunge With Movement of the Finger

Fixed point: head; mobile point: finger.
Perform a side lunge with your finger in front of your face. In this position, move your arm to the left and follow the finger only with your eyes.

Movement: Side Lunge With Opposing Movement of the Head

Fixed point: finger; mobile point: head.
As you perform the side lunge, keep your finger still and turn your head to the right, keeping your eyes on your finger.

Wow, what coordination!

Work On

- Stimulating the nervous system.
- Stimulating coordination.
- Stimulating proprioception.
- Reducing tension in the neck.
- Improving vision.

Discussion

These are exercises requiring great coordination, and it is a shame they are not more widely known.

The eyes are optical instruments (to see, to transmit light that the brain processes as images), and they are also structures equipped with muscles that point and guide the eyes in space. This means that the eyes can both be considered from an optical standpoint (i.e., how well they see when examined by an optician), as well as an oculomotor standpoint (how they move).

ON/OFF Shortened Hip Flexor

Starting Position

Stand up straight with your legs hip-width apart and your arms straight and down by your side.

Focus On

- Active kite.
- Fixed point: hand; mobile point: knee.

Work On

ON/OFF = activation in shortened hip flexor and relaxation.

Discussion

Why is the hand only providing resistance? The aim is to isometrically work the shortened hip flexor. That is why the hand is the "conductor" of the movement; otherwise, focus would be lost.

Movement

Raise your right leg in 90–90–90 degree Stork pose (feet, knee, and hip bent at 90 degrees). Place your right hand on your right knee. Push your knee against your hand and resist for 3 seconds. Return to the starting position and relax the hip flexor for 3 seconds. Repeat 10 times and then switch legs.

Change Stimulus

Lie flat on your back and bend and raise your right leg to hip height. There should be a 90-degree angle between your thigh and leg and a 90-degree angle between your torso and thigh. Place your left hand on your right knee. Slightly rotate your thigh outward, moving your foot inward. Press your knee against your hand for 3 seconds then lower your right leg to the ground and relax for 3 seconds. Repeat 10 times and then change sides.

Strategy
FEEL IT – ON/OFF Activation

Starting Position
Adopt a half-kneeling position, with your left foot out to your left side. Turn your right leg inward. Place your hand on the inside of your left knee.

Movement
Move your left knee toward your toes: Your pelvis will lower slightly, thereby increasing the stretch of the inner thigh and of the hip flexor. Push your knee against your hand and hold for 3 seconds. Relax and return to the starting position for another 3 seconds. Repeat 10 times and then change sides.

Focus On
- Active kite.
- Fixed point: hand; mobile point: knee.

Work On
ON/OFF = isometric activation in stretching of the hip flexor and relaxation.

Discussion
Very few people are completely symmetrical and perfectly aligned. In this exercise it is very common to notice a difference between the left and right legs; one hip feels freer than the other. It is the range and fluidity of movement that change.

Beat (Have Fun Tapping)

Starting Position and Movement

Adopt a half-kneeling position with your left leg forward. Start to tap your whole leg with your hands and then move to the thigh. Next, tap the side of your rib cage with your right hand before moving to the left arm (outer and inner arm). Continue to tap your left arm down to your hand and then start vigorously clapping. Are you awake?!

The entire exercise can also be performed standing up.

Focus On

The action is a light tap to stimulate but not to leave marks or bruises.

Work On

- Systemic activation.
- Activating the sympathetic autonomic nervous system.
- Waking up the mechanoreceptors.
- Movement preparation.

Discussion

This seemingly silly exercise is in fact fun and effective. It is much loved by my clients because it can elicit a smile in an instant.

Include the miraculous sequence detailed in the before and after test below in your next warm-up.

Test

The before and after test consists of performing the three exercises shown here (see pages 90, 91, and 92) but only on the right leg.

After 10 repetitions of each exercise, perform a leg bend, a walk, a side lunge (see page 88), and a squat movement. Repeat from the beginning with the other leg.

Strategy

FEEL IT – Activation

Starting Position and Movement

You can perform this exercise in any of the starting positions. Completely open and then completely close your hand. Start slowly before increasing the speed. Repeat 10 to 20 times and then change hands.

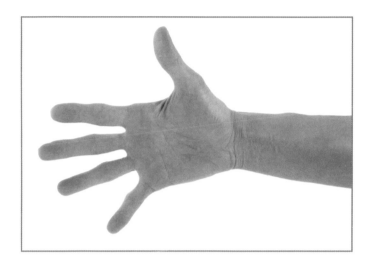

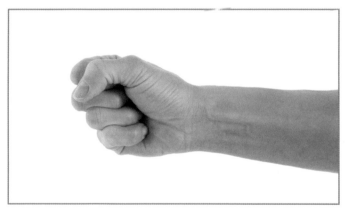

Work On

Neuro-myofascial activation.

4.3.3. FEEL IT Exercises With Equipment

Anterior Stimulation of the Ankles Resting on the Foam Roller (SFL)

Strategy
FEEL IT – SFL Activation

Starting Position

Sit on both your heels and place the foam roller between the floor and your ankles so that your feet can move freely. Ensure that the spinal column is in a neutral position, with an active kite.

Movement

With slow and fluid movements, bend and extend your ankles by pressing them down gently on the foam roller. Repeat 10 times.

Focus On

- Fluid movements.
- Maintaining constant gentle pressure on the foam roller.

Work On

- Relaxing the anterior tissue of the ankle (the superior and inferior retinaculum of the tendons of the toe extensors, SFL).
- Increasing the mobility of the ankle.
- Improving gait.
- Improving squats (increasing squat ROM).

Discussion

Despite being a very simple exercise, it has a huge impact on ankle mobility. How often do you meet clients with an immobile ankle joint that hinders their ability to perform leg-bending exercises?

The problem concerns the ligaments of the instep, which are also known as the superior and inferior extensor retinaculum. To explain it in simple terms, you can think of a retinaculum as a restraint fascia that surrounds the ankle. This fascia holds the tendons of the SFL (extensor digitorum longus and brevis, tibialis anterior, anterior crural compartment) in place and directs their force from the tibialis muscle to the toes, acting as a pulley. Imagine if this band of restraint around the ankle becomes stiff, inhibited, and does not allow the underlying tendons to slide, thereby hindering the movement of the ankle and toes. What could cause this? A sedentary lifestyle, incorrect movements, accidents, and inappropriate shoes are just some possibilities.

The aim of this exercise is to free up the tendons by rehydrating and stimulating this area. It is an easy way to reduce tension in the anterior part of the ankle. You can find everything you need to know about how to use foam rollers and balls correctly on page 263.

Change Stimulus

- Use sensory balls (with soft spikes: We don't want it to hurt).
- If you don't have a foam roller, use a rolled up towel.

Increase the Level of Difficulty

Perform the same movement with only your right ankle on the foam roller. What's changed? There is increased load on a single ankle.

Test

I often meet people who are unable to kneel on their heels (feet flat). This may be because the SFL is too tight. Try this before and after test:

1. Sit on your heels, with your feet flat (without help). Make a mental note of the tension felt and how close your buttocks come to your heels.
2. Perform the exercise with the foam roller or with two balls.
3. Sit on your heels again, with your feet flat (without help). Make a mental note of the tension felt and how close you are to sitting down.

Isn't the result amazing?

Posterior Stimulation of the Ankles Resting on the Foam Roller (SBL)

Strategy
FEEL IT – SBL Activation

Starting Position and Movement

Kneel on the floor with your feet flat and place the foam roller between your ankles and calf muscle. Slowly sit down on the foam roller and apply gentle pressure for 3 seconds, then lift your buttocks to release the tension. Slowly sit back down on the foam roller and hold the position for a few seconds, increasing your sitting time. Keep your back in a neutral position, with active kite. Repeat 5 to 10 times.

Focus On

No pain, no gain? This is not our focus. Adjust the intensity to the tension you feel.

Work On

Relaxation of the posterior ankle tissue (calf muscle, soleus, SBL).

Pressure and Massage on the Calf Muscle With the Foam Roller (SBL)

Strategy

FEEL IT – SBL Activation

Starting Position and Movement

Sit down with your hands resting comfortably on the floor behind your chest, your back straight, and with active kite. Rest your right leg on the foam roller between your ankle and calf muscle. Use your left foot to stop the foam roller from moving. Gently push your calf muscle down onto the foam roller for 3 seconds. Repeat 10 times.

Focus On

Appropriate, unforced pressure.

Work On

- Stimulating the leg tissue/SBL.
- Eliminating tension.
- Increasing the work capacity of the lower limbs.

Change Stimulus

- Keep pressing down gently and carefully move your leg left and right, massaging the tissue for another 10 seconds. Switch legs.
- Change the position of your foot (internal rotation, external rotation) as you press down.

Discussion

The foam roller does not have a correct position per se, but if you move it slightly toward your ankle or toward your knee, you will find the correct tension for you. Once you have found the correct position, keep it there throughout the exercise.

This exercise relaxes and activates a part of the SBL (the Achilles tendon, which connects the calf muscle to the calcaneus), to improve the fluidity of movement of the ankle and knee.

Strategy
FEEL IT – DFL Activation

Starting Position and Movement

Lie down sideways with your weight on your right elbow, your back straight, and with active kite. Place the foam roller between the ankle and calf muscle of your straight left leg and place your right leg in a comfortable position on the floor. The foam roller does not have to be put in a specific position, but if you move it slightly toward your ankle or toward your knee, or slightly rotate your leg inward or outward, you will find the correct tension for you. Once you have found the correct position, keep it there throughout the exercise.

Gently push your left leg down onto the foam roller for 3 seconds. Repeat 10 times. Then slowly move your left leg back and forth over the foam roller for a further 10 seconds, massaging the tissue. Switch legs.

Focus On

Active kite.

Work On

- Stimulating the leg tissue/DFL.
- Eliminating tension.
- Increasing the work capacity of the lower limbs.

Pressure and Massage on the Inner Thigh With the Foam Roller (DFL)

Strategy
FEEL IT – DFL Activation

Starting Position and Movement

Lie down on your front with your weight on your elbows, your back straight, and with active kite. Place the foam roller under your left inner thigh, close to your left knee. Stretch your right leg out comfortably on the floor. The foam roller does not have to be put in a specific position, but if you move it slightly toward your knee or toward your pelvis, or slightly rotate your thigh inward, you will find the correct tension for you. Once you have found the correct position, keep it there throughout the exercise.

Gently push your left leg down onto the foam roller for 3 seconds. Repeat 10 times. Then slowly move your left leg back and forth over the foam roller for a further 10 seconds, massaging the tissue. Switch legs.

Focus On

Active kite.

Work On

- Stimulating the leg tissue/DFL.
- Eliminating tension.
- Increasing the work capacity of the lower limbs.

Change Stimulus

Perform the same exercise, but this time place the foam roller under your inner thigh closer to your pelvis.

Crunches on Sensory Balls (SBL/DFL)

Starting Position

Lie down on your back with your legs bent and hip-width apart. Place the sensory balls on either side of the spine, below the shoulder blades. Interlock your fingers and place your hands on the nape of your neck.

Movement

Breathe out and lift your chest upward, looking at your knees. Breathe in and return to the starting position. Repeat 5 times. Then move the balls about half an inch (1 cm) toward your pelvis. Repeat 5 times.

Focus On

- Active kite.
- Steady breathing.

Work On

- Activating the sympathetic autonomic nervous system.
- Stimulating the SBL/DFL tissue.
- Releasing tension in the thoracic spine.
- Improving breathing.
- Increasing mobility and flexibility.
- Strengthening the core muscles.

Discussion

Try combining this exercise with the Breathalyzer Breathing exercise (page 86).
Raise your chest in 3 phases as you exhale 3 consecutive times. Expel all remaining air from your lungs with your third exhale. Do not breathe in between the three exhalations. Try to make a "CHHH" sound as you breathe out.

Strategy
FEEL IT – LL Activation

Starting Position

Stand up straight with your hands by your sides and the mini band around the middle of both your feet. Space your legs far enough apart to slightly stretch the mini band. Point your toes forward.

Focus On

- Keeping your knees straight.
- Taking the side step by pushing on the mini band with the outside of your foot.
- Active kite.

Work On

- Activating the lower part of the LL.
- Activating the leg stabilizers.
- Protecting the knees.

Reduce the Level of Difficulty

- To start, use a looser mini band.
- Perform the whole exercise without a mini band (also see pages 120 and 121).

Change Stimulus

- Perform the side steps with your feet turned inward. This will stimulate the LL in a stretched position of tension.
- Perform the side steps with your feet turned outward. This will stimulate the LL in a shortened position of tension.

Movement

Take a small step sideways with your left foot. Move your right leg toward your left, ensuring that the mini band is always slightly stretched. Continue until you have completed 5 side steps. Return to the starting point and repeat the exercise with the right leg.

Discussion

I first saw this exercise in 2009 performed by Stuart McGill, PhD, a leader in his field and Professor of Spine Biomechanics at the University of Waterloo in Canada. I have been hooked ever since.

I have interpreted it, working with the myofascial lines as follows: Choosing to place the mini band around the middle of the feet activates the lower part of the LL, which passes through the middle of the foot at the base of the first and fifth metatarsal (also see the discussion on page 121).

MOBILITY: FUNCTIONAL MOBILITY

MOBILITY

MOBILITY is a strategy that combines scientific methods with doable exercises to actively maximize joint range of motion.
Remember: **Our joints betray our age**.

5.1. MOBILITY STRATEGY

In this chapter we will look in detail at why and how I came to the decision to include the MOBILITY strategy (functional and joint mobility) in my training.

As we have already said, the FReE method strategies follow a logical order, but the idea is not to follow a default plan because no two people are the same. What is important is understanding why we make certain choices. It is important to know what to do, how to do it and why. Choosing to perform functional mobility exercises has many benefits.

The MOBILITY training that I propose in this book is varied, involves the primary joints, and uses a systematic approach to achieve functional mobility. As such, it offers the following benefits:

- It improves joint mobility.
- It maintains and improves joint functionality.
- It improves the functionality of everyday movements.
- It stimulates the nervous system.
- It keeps the joints young.
- It lubricates and nourishes the joints.
- It prevents and treats osteoarthritis.
- It improves athletic performance.
- It improves muscle flexibility.
- It communicates with the brain and "feeds the mind".

In a nutshell, joint mobility is vital for total well-being and athletic performance, which is why continuous and systematic training is required. Unfortunately, this is very often overlooked because it is considered to be boring or a waste of time.

Let's take a closer look at what functional or joint mobility actually means.

Definition of Joint Mobility

Joint mobility is the ability of a joint or articular system to move with the greatest possible *range of motion* (ROM).

Factors that influence joint mobility are as follows:

- Structure of the joint
- Muscle and tendon elasticity
- Ambient temperature
- Warm-up and body temperature
- Age and sex
- Anxiety and stress
- Level of muscle fatigue (this limits the action of agonist and antagonist muscles)

Classification of Joint Mobility

Joint mobility is divided into the following categories:

- *Anatomical joint mobility*: Range of motion of the joint permitted by the anatomical nature of the limiting components (facet joints, extensibility of connective and muscular structures).
- *Active joint mobility*: Maximum range of motion of a joint by contracting the agonist muscles and, at the same time, relaxing the antagonist muscles. It is therefore influenced by the level of muscle strength and extensibility.
- *Passive joint mobility*: Maximum range of motion that can be achieved with the help of external forces (a training partner or equipment). It is based on the ability of antagonist muscles to relax and stretch. It is therefore influenced by potential loading forces by a partner, or by muscle extensibility.

Passive joint mobility is always greater than active mobility and should account for 90 percent of anatomical mobility. The difference between passive and active joint mobility represents the extent to which active mobility can improve, either by enhancing the agonists or increasing the antagonists' ability to stretch.

The ability of a muscle to stretch during a joint movement is known as muscle flexibility, which can be improved with training. However, **muscle flexibility** may be limited by the joint capsule, the activity of the contractile component of the muscle, the connective tissue of the muscle itself, and its tendons, as well as the skin. Combining joint mobility with muscle flexibility leads to a discussion on muscle-joint flexibility, which will be addressed in chapter 6.

Resistance to Stretching

People used to believe that mobility is limited by the muscles themselves, which must therefore be trained. However, it is now known that connective tissue (and not just the muscles) around the joints (capsules, ligaments) and the fascial tissue of the muscle are responsible for limiting joint movement [25; 52].

There is a reason for this. Exceeding the physiological range of motion of a joint tends to damage the structure of the capsules and ligaments but does not cause the muscles surrounding the joints to tear. Muscles are usually injured at normal joint angles. Studies show that muscle injuries are associated with a lack of coordination and may have nothing to do with a lack of mobility.

This discovery has, and will continue to have, a significant impact on the planning of stretching and strength courses and on prevention.

In percentage terms, which elements hinder mobility the most?

- Joint capsule: 47 percent
- Antagonists and fascia: 41 percent
- Tendons and ligaments: 10 percent
- Skin: 2 percent

Benefits of Joint Mobility

Functional mobility is essentially a dynamic exercise that teaches the body how to work within different ranges and therefore how to control the movements themselves. Why is it important to mobilize the joints?

1) To lubricate and protect the cartilage (mechanical function)
2) To give instant feedback to the central nervous system (neurological function)
3) To increase or maintain the ROM
4) Because every joint involves multiple myofascial lines

1) Joint Lubrication

Including the joints in your warm-up gives them the chance to lubricate. Regularly mobilizing the joints and subjecting them to appropriate loads may help to slow down the aging process.

The joint surfaces (joint cavity and head) are covered with a layer of cartilage, which acts as an airbag to protect the joints. The cartilage, which supports the joint, covers the ends of the bones in order to protect the joints from the friction that occurs during movement. It is extremely resistant to traction/pressure and is able to deform. Cartilage is made up of cartilage cells (chondrocytes), connective tissue fibers (collagen), and an impermeable substance. The chondrocytes form groups (chondrons) bound by connective tissue fibers, which in turn create microscopic pads that help to protect against and cushion the compressive loads on the cartilage [46]. A distinctive feature of cartilage is that it does not contain any blood capillaries. It is supplied with nutrients and oxygen through joint lubrication (movement) produced by the so-called synovial membrane (inner layer of the joint capsule made up of elastic fibers and fat) in the intra-articular space when it is in motion. After that, it penetrates the cartilage layer.

Remember that it is better to prevent osteoarthritis than to treat it.

Negative notes: The cartilage of a joint that is not sufficiently mobilized or loaded will not receive enough fluid, causing its thickness, elasticity, and resistance to decrease. At the same time, cartilage may be damaged by excessive strain and inappropriate loads.

2) Instant Feedback

The functional mobility exercises have an instant impact on the central nervous system. In section 1.8 we discussed the numerous mechanoreceptors that are found around and within the muscles and joints, in the tendons, in the ligaments, and in the capsules; these mechanoreceptors transmit information to the central nervous system. Joint mobilization takes advantage of this instant communication. The feet, hands, and back contain multiple joints. We use this collection of receptors to communicate more quickly with our CNS and to obtain a magnified response.

3) Increasing or Maintaining ROM
This important aspect was covered at the beginning of the chapter.

4) Coordination of the Myofascial Lines
Multiple myofascial lines pass through every joint in our body. The synergy between the individual myofascial lines is coordinated in part by joint mobilization.

Controlled Mobilization
The FReE method uses controlled mobilization to increase, control, and create:

- *Increase*: Strengthen and increase the *range of motion** of the joint.
- *Control*: Recruit the nervous system to optimize the active *range of motion* of the limb; progressively induce tissue adaptation to preserve it.
- *Create*: Develop the fluidity of the basic movement, gradually performing more complex movements to enhance the movement flow.

* *Range of motion* (ROM) is the freedom of movement that a joint can physiologically perform, expressed in degrees.

Controlled mobility

- increases control of short, medium, and long *range of motion* (intermediate range of motion is normally controlled),
- is the quickest way to increase flexibility,
- is integrated into functional movement training,

- gives rise to agility (the ability to effectively change body position), and
- increases joint stability (because it increases our neurological control).

Movement is not possible without mobility.

Have you ever wondered which ranges of motion people use during a typical day? Most people only use 40 to 50 percent of the possible range (for a variety of reasons that have already been mentioned). Then they begin to train, with the aim of using the complete range of motion in the exercises. Can it be done? Yes, but only with compensations.

It is now clear that the right conditions must be created, and the foundations or "invisible bricks" on which to build everything else must be laid, before attempting an exercise that requires a wide range of motion.

5.2. JOINT BY JOINT THEORY

The first time I ever came across joint mobility tests was back in 2001 when I was taking part as Reebok Master Trainer in the Reactive Neuromuscular Training: 5 Point Movement Screen course at Reebok University in Boston, run by Gray Cook. It was a method light years ahead of its time. When I proposed it in Switzerland for the first time in 2002, it did not receive a great response. I didn't let that stop me, and I have constantly included it in my courses. Today, reading our joint systems, using a method proposed and illustrated by the great Gray Cook and later picked up by Michael Boyle, the creator of functional training in America, is a fundamental part of our training programs.

Joint by joint theory is the key to reading our joint systems, with significant benefits in terms of functional recovery for rehabilitation, as well as prevention.

Put simply, this theory highlights a number of important things:

1) The joints alternate between mobility and stability.
2) What tends to happen is that when we experience muscle problems, tension, or an inflamed or painful joint (unless pathological or genetic), the real problem is typically not there but in the joint below.
3) The only exception is the hip.

Joint	Characteristics
Tibiotarsal	Mobility
Knee	Stability
Coxofemoral	Mobility
Lumbar spine	Stability
Thoracic sternum	Mobility
Scapula	Stability
Glenohumeral	Mobility
Elbow	Stability
Wrist	Mobility

Our body is a single unit: We divide it into its anatomical parts, endogenous systems, and muscles, even in training, but the body doesn't know that and works as a unit.

None of our routine movements is made in isolation; we do not work with a single joint at a time. We work with muscle chains (myofascial lines) and joint chains.

5.2.1. Mobility and Stability

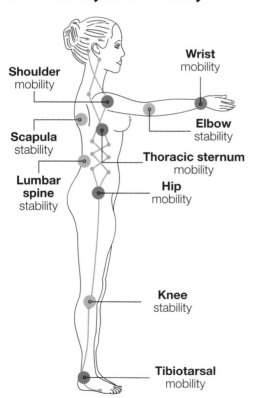

Wrist mobility

Shoulder mobility

Elbow stability

Scapula stability

Thoracic sternum mobility

Lumbar spine stability

Hip mobility

Knee stability

Tibiotarsal mobility

5.2.2. Types of Movement

The main types of joint movement are as follows:

1) Gliding
2) Rolling
3) Combined rolling and gliding
4) Axial rotation

From a functional and biomechanical perspective, it is not the range of motion (ROM) of each individual joint that is so important but rather the effect resulting from the combination of the movements performed by the joints arranged in succession.

5.3. MOBILITY STRATEGY PLAN

When should mobilization be included? Given all the positive aspects that I described at the beginning of the chapter, it is important to include mobilization at the beginning of a training session or in the morning to start the day well. A systemic warm-up prior to mobilization may be useful in a training session to "turn your engine on" each morning. Start with small movements before gradually increasing the intensity.

Repetitions
Repeat each exercise 2 to 10 times.

Variability
Did you know that joint mobility varies throughout the day? It is lower in the morning and increases throughout the day as the temperature increases. In other words, joint mobility is lower in the cold and higher in the warmth. In training, it increases as we warm up and decreases after a strenuous workout.

5.4. CATEGORIZATION OF FUNCTIONAL MOBILITY EXERCISES

In this chapter, the functional mobility exercises are grouped by joint, progression, and equipment.

- **Body weight exercises**
  Functional mobility: ankle/foot
  Functional mobility: hip
  Functional mobility: spinal column
  Functional mobility: scapulohumeral joint

- **Exercises against the wall**
  Functional mobility: hip
  Functional mobility: hip and shoulder
  Functional mobility: scapulohumeral joint
  Functional mobility: spinal column

- **Exercises with equipment**
  Functional mobility: hip
  Functional mobility: scapulohumeral joint

5.4.1. MOBILITY Body Weight Exercises

"Turn on your engine" with the following high-impact proprioceptive exercises.

5.4.1.1. Functional Mobility: Ankle/Foot

Positions of the Foot			
Pronation			
Correct posture			
Supination			

The goal of the following exercises is to reset foot posture by correcting incorrect foot positions like excessive pronation or supination.

Strategy

MOBILITY – Functional mobility – Ankle and foot

Starting Position

Stand up straight, with your feet hip-width apart and your arms down by your side.

Movement

Take one step forward with your right leg, keeping your left leg straight and your left foot (3 points of contact) on the floor.

From this position take a step backward, again with your right leg. As you step backward you should feel the lower part of the SFL stretch (at the thigh = quadriceps).

Repeat this exercise (back and forth) with your right leg 10 times and then switch to the left leg. Move your arms as you would when walking and keep your head facing forward.

Focus On

- The three points of contact of the foot (in this case the left foot) should not move. The heel is also part of the foot and must not lift off the ground.
- The left leg (in this example) must always be straight.

Work On

- Functional mobilization of the tibiotarsal joint on the sagittal and transverse planes.
- Mobilizing the hip.
- Dynamic stretching and activation of the SBL/SFL/LL/DFL.

Change Stimulus

- Perform the entire exercise with your arms outstretched above your head. This will increase the stretching and activation of the upper part of the SFL (abdominals). Further increase the stretch by extending your chest and torso.

- "L" step: Reduce the length of the forward step. As you move your right foot forward, turn your toes inward, forming an "L" shape with your feet. Your right heel should be in line with the toes of your left foot. Start with your hands down by your side. If you wish, perform the exercise again with your arms raised above your head.

- Return to the starting position (standing up straight), raise your arms in front of your chest to shoulder height, and step back and forth with your right leg. As you step forward, turn your chest to the right and bend your right elbow. Do not move your left arm, and keep your head facing forward. As you step backward, move your right arm forward to the starting position. Repeat another 5 to 8 times, and then change sides. This mobilizes the thoracic spine, recreating the rotation that normally occurs when walking. Is it functional? I would say so.

- Use a resistance band to help you keep the three points of contact of your foot firmly on the floor.

Discussion

This exercise is fantastic because, despite its simplicity, it is extremely effective at mobilizing the tibiotarsal joint. On the surface it may seem simple, pointless, or laughable. But its true effectiveness lies in its simplicity. Don't believe me? Try the following test.

Test

Perform the entire exercise exactly as described above but this time only on one side.

This time, only move your right foot while keeping your left foot stationary. Perform the exercise 10 to 15 times, either with or without the "L" step.

Once you have finished, it's time to see whether there is any difference in mobility between your right and left ankles. Listen to the changes of your body:

1. Raise your right leg without losing your balance (Stork pose); repeat with your left leg. Do you notice any difference in stability or agility?
2. In the Stork pose and with your left foot on the floor, raise your left heel, and then change sides. Do you notice any change of mobility?
3. Perform a single-leg squat first on one leg and then on the other. Do you notice any difference in mobility and freedom of movement?

Anyone who is able to stand can perform this exercise.

Strategy

MOBILITY – Functional mobility – Ankle and foot

Starting Position

Stand up straight, with your feet hip-width apart and your arms down by your sides.

Movement

Take one step sideways with your right leg, keeping your left leg straight and your left foot fully on the floor and facing forward.

Push off firmly with your right leg to return to the starting position (or go directly with your right foot forward), and turn your toes inward, forming an "L" shape with your feet. Repeat as desired.

Now take a diagonal step backward on the transverse plane, keeping your left leg straight and your foot firmly on the floor and pointing forward. Return to the neutral or "L" starting position.

Focus On

- Fixed point: foot that is on the floor; mobile point: foot that moves sideways.
- In this case, the left foot is always in contact with the floor, with the toes pointing forward; the left leg must be straight.
- Active kite.

Work On

- Functional mobilization of the tibiotarsal joint on the coronal and transverse planes.
- Dynamic stretching and activation of the SFL/LL/DFL/SL.
- Strengthening the muscles of the lower limb

Discussion

This is a more advanced version of the "Stepping Back and Forth" exercise (page 109) but on different planes. The different position of the leg during the movements affects and partly changes the myofascial lines involved. The focus is still on mobilizing the tibiotarsal joint, including movement on the coronal and transverse plane; increasing the level of exertion increases mobilization of the hip joint. By moving sideways, you should feel your adductors stretching more. From a myofascial perspective, the Deep Front Line (DFL) is involved. In transverse movement, the adductors still stretch, but the hamstrings, which are part of the SBL, are also involved.

Increase the Level of Difficulty

Side lunge: Take a bigger step to your right and bend your hip, moving it slightly behind your buttocks. Keep your back in a neutral position and as straight as possible. Bend your right leg, touching the floor with the fingertips of your right hand; your arm should pass on the inside of your right knee. Push off firmly with your right foot to return to the starting position, or go directly to the "L" position with your right foot. Start with smaller steps before gradually increasing the ROM. Repeat 5 to 10 times and then change sides.

Change Stimulus

Diagonal lunge: Perform exactly the same exercise, but this time take a diagonal step backward. Start with smaller steps before gradually increasing the ROM. Repeat 5 to 10 times and then change sides.

Discussion

These exercises have a very significant impact on ankle and hip mobility. Additionally, moving in all directions improves the sliding of the various myofascial layers (LL/DFL/SBL/SFL/SL and adductors, abductors, hamstrings, quadriceps, pelvic floor), thereby relieving tension. All these myofascial lines are indirectly linked to the pelvis.

Think of your pelvis as a tent. The bones are the central supports, and the myofascial components are the guy ropes. When putting up the tent, if the guy ropes are not evenly distributed around the structure, the tent will not be correctly balanced; in other words, it will be askew. The same is true of the pelvis, where the excessive tension of certain myofascial lines may cause an imbalance. Change the tension and the bones return to their correct position. I include exercises like "Stepping Back and Forth" and "Side Step" in my warm-up routines to balance the tension and facilitate the sliding of the different layers. Outcome: The training that follows will be more effective. Of course we cannot perform miracles. If you suffer from a particular condition, treatment will be more tailored from specialists, such as chiropractors, osteopaths, and physiotherapists.

See chapter 9 for some specific programs.

Sagittal Woodpecker: Movements in Half-Kneeling Position

Strategy
MOBILITY – Functional mobility – Ankle joint

Starting Position

Adopt a half-kneeling position: Left leg forward, right knee on the floor, foot in hammer position, chest tilting forward slightly, hands resting on your thigh.

Movement

Using all your body weight, move your left knee forward, past your toes. Do not let your left heel lift off the floor.

Repeat this exercise (moving your body weight backward and forward) with your left leg 10 times, and then switch to the right leg.

Focus On

- Fixed point: front foot; mobile point: knee.
- Moving your knee as far past your toes as possible.
- Keeping your heel on the floor at all times.

Work On

- Functional mobility of the ankle.
- Multiple strategies and multiple myofascial lines are involved.

Progressions on the Coronal/Transverse Plane

Continue moving your knee but changing the direction. Move your front knee outward, and then return to the middle.

Move your front knee inward and then return to the middle. Finally, move your front knee in circles around the ankle clockwise and counter-clockwise.

Discussion

Improving the tibiotarsal joint requires hard work. What do I mean by that? You will have to exert yourself to go beyond your toes. The joint should not become painful, but it should be stimulated. Stimulate is synonymous with change, improve...

Frontal Woodpecker: Movements in Half-Kneeling Position

Strategy

MOBILITY – Functional mobility – Ankle joint

Starting Position and Movement

Adopt a half-kneeling position: Move your left foot outward, with your left heel in line with your right knee and with your left hand resting on your left knee. In this position, perform all the movements described in the previous exercises (moving forward, inward, outward, and in circles).

Discussion

This is the natural progression of the previous exercise ("Sagittal Woodpecker"); everything that applies to the previous exercise also applies to this one, so there is no need to repeat it here.

I will add one thing: Don't place your left foot too far from your right knee, or you will be stretching your inner thigh rather than mobilizing the tibiotarsal joint, which is our goal.

Strategy

MOBILITY – Functional mobility – Tibiotarsal joint, foot, and hip

Starting Position

Adopt a half-kneeling position, as described in the previous exercise. This time, however, sit on your right heel, as the name of the exercise suggests. Let me clarify: Place your left foot flat on the ground in front of your torso with your left knee forward; rest the forefoot of your right foot on the ground with toes bent and with your right knee on the ground.

Keep your chest and torso straight and erect. Place your hands on your knees.

Movement

With a fluid motion, move both your knees apart and outward. Keep your left foot and your right forefoot firmly on the ground.

Benefits of All This Pain

- Less painful and more agile feet.
- More fluid gait.
- Improved foot support.
- Improved hallux valgus.

Focus On

- Fixed point: foot; mobile point: knee.
- Performing fluid movements.
- You shouldn't feel too much pain. The exercise is counterproductive if it is too intense; do a simpler exercise if necessary.

Work On

- Mobilizing the tibiotarsal joint, metatarsophalangeal joint, and hip.
- Stretching the plantar fascia and flexor digitorum brevis at the same time.
- Involving multiple strategies and multiple myofascial lines.
- Strengthening the muscles of the lower limbs.

Discussion

This is a progression of the previous exercises to mobilize the tibiotarsal joint and the hip joint. In this position a good deal of mobility and flexibility is required. Proceed with caution, or you will do yourself more harm than good.

At first I found this exercise really difficult, but I got the hang of it after a few attempts. Take your time.

P.S.: For women, think of an evening in high heels, when your feet are crying out for revenge. That is the right time to perform this graceful movement.

VARIATION: Goose Walk in Half-Sitting Position

From the starting position described in the previous exercise, move your right leg forward and lower your left knee toward the floor; continue moving forward and backward in this way.

Focus On

- Active kite.
- Keeping your pelvis low when walking.
- Knees facing your toes.

Work On

- Ankle and foot mobility.
- Strengthening the leg muscles, particularly the quadriceps.
- Asymmetrical coordination.

Discussion

I learned this exercise many years ago when I was doing boxing training. In all honestly, I couldn't stand it because it was really difficult and I felt silly: It involved walking in circles like a goose.

This exercise is not suitable for everyone because it is a progression of all the other exercises that we have already seen.

In terms of its benefits, it is great for athletes because most lack strength in a full squat.

Strategy
MOBILITY – Functional mobility – Joints of the foot

Starting Position
Squat, with your feet slightly more than shoulder-width apart. Keep your back straight and in a neutral position. Put your hands together and push your elbows gently against your inner thighs (giving your knees input to open outward).

Movement
Shift your body weight to your left forefoot (precisely on the first and fifth metatarsal head) by raising your heel off the ground. Return to the starting position and repeat on the right side. Continue shifting from one side to the other.

If you are able to hold this position without overexerting yourself, stay on the left side, and move your knee in circles around your forefoot.

Move your big toe away from your second toe. Repeat a few times and then change direction. Repeat with the other foot.

Focus On
*   Moving your knee in circles around your forefoot.
*   Moving your big toe away from your second toe.

Work On
*   Involving multiple myofascial lines.
*   Strengthening the muscles of the lower limbs.
*   Mobilizing the joints of the toes, metatarsal heads, and phalanges of the foot.

Discussion
Creating space between your first and second metatarsal head will help you walk. When the back foot is moving forward, the forefoot widens, giving us the thrust needed to walk (between the first and second metatarsal heads). This stimulates the interosseous muscles, which has an impact on the quadriceps. What happens? As the metatarsals extend, the mechanoreceptors in the tissue between the metatarsals record the movement and transmit the impulse to the knee extensors, which absorb the force of gravity and the reaction force on the ground. That is why the space between the first and second metatarsals is important.

Runner's Knee in Squatting Position

Strategy
MOBILITY – Functional mobility – Hip, foot and ankle joints

Starting Position

Adopt the same squat position as described in the previous exercise.

Focus On

- Moving your knees away from each other.
- The foot pointing outward must remain flat on the ground.
- Keeping your back as straight as possible.

Work On

- Mobility of the hip, foot, and ankle joints.
- Stretching the inner thigh (DFL).
- Muscle strengthening.

Movement

Place your left knee on the floor in front of you, leaving your right leg and foot in the starting position. At the same time, gently push your right elbow against your inner right thigh, stimulating your leg by moving it backward to increase the ROM.

Return to the starting position to repeat the movement on the other side.

Discussion

By working on different components and planes, an exercise becomes more dynamic, resulting in a loss of control from a technical perspective. This is not my intention.

Remember that fascia needs time to adapt to a new situation. Begin with slow, controlled movements. Once you have learned the correct technique, start to change the rhythm: Increase the speed – reduce the speed – reduce and increase the speed. Our movements, and their time/rhythm in particular, should be as versatile as the fascia itself.

Moving Stork

Strategy
MOBILITY – Functional mobility – Hip joint

Starting Position
Adopt the 90–90 degree Stork position, with your right knee raised to hip height, forming a 90-degree angle between your hip and your torso and 90-degree abduction between your lower and upper leg. Raise your outstretched arms to shoulder height.

Movement
Rotate your pelvis around the head of your left femur, moving your right leg outward as you do so but keeping your shoulders and left leg still. Return to the starting position. Repeat 10 times and then change legs.

Focus On
- Fixed points: shoulders and standing leg; mobile point: pelvis.
- Active kite.
- Stretch from your standing heel to the top of your head: This will keep you standing upright and stop your left side from moving too much.

Work On
- Functional mobility of the hip joint.
- Coordination.
- Balance.

Discussion
It is not easy to do such selective exercises in a position where you could easily lose your balance.

If you feel unstable, hold onto or lean against something with your hand.

Imagine that you are screwing or unscrewing the top of a thermos: Hold the base still (like the base of your body) and screw on the lid (your leg in the air); this is the same movement that we do in this exercise.

Knee Saver – Lateral Stabilization

Strategy
MOBILITY – Functional mobility – Hip joint

Starting Position

Lie on your left side with your left forearm on the floor and your left shoulder perpendicular to your elbow. Place your legs and feet on top of each other. Align your heels, gluteal muscles, and shoulders: Imagine that you are leaning against a wall.

Movement

Before the movement, create pre-tension on the lateral line (LL). Push your left shoulder blade/shoulder toward your legs, pulling your elbow slightly toward your pelvis. You should feel the left side of your torso activate (oblique muscles, intercostals, serratus anterior). Push the outside of your left foot and knee gently into the floor. Finally, raise your pelvis upward and lift your right leg to hip height, keeping it bent.

Lower your pelvis until it touches the floor before lifting it again. Complete the desired number of repetitions and repeat on the other side.

Focus On

*   Keeping your foot on the ground.
*   Maintaining the points of contact with the ground: Outer edge of your foot, knee, and elbow.
*   The push is provided by the outer edge of your foot and knee, keeping your heel as close to the floor as possible. It is precisely this position of the foot (due to postural imbalances) that often makes this exercise difficult but at the same time effective.
*   Rhythm: Lower your pelvis slowly, and raise it in an elastic but controlled manner. Put the lateral line under tension as you go down. Use this pre-tension to lift your pelvis back to its raised position.

Change Stimulus

- Adopt the starting position as described in the previous exercise, but this time extend your right leg. Raise and lower your pelvis, complete the desired number of repetitions and repeat on the other side.

- This movement will mobilize the spinal column. Adopt the starting position of the previous exercise, with your pelvis raised. Turn your torso in a controlled movement, bringing your right arm under your side in the gap between your torso and the floor. Keep your pelvis still. To simplify the exercise, keep your right foot on the ground, or you can adopt the starting position with both your legs bent, one on top of the other. Complete the desired number of repetitions and repeat on the other side.

- Adopt the starting position as described in the previous exercise, but this time place your right foot in front of your left knee. Place your hand on the floor instead of your elbow. As you raise your pelvis, push your right heel into the floor (this will activate the back of the thigh), with your forefoot firmly on the ground. Keep your pelvis perpendicular to the ground and return to the starting position. Complete the desired number of repetitions and repeat on the other side.

Discussion

Guido Bruscia calls this exercise the "knee saver", and it really is.

It is fantastic and has a genuine *wow* factor. I often meet clients who have bad knees for a variety of reasons. As a personal trainer, my job is to help my clients improve their strength and fitness while preventing injury. Many of my clients have knee pain even before performing a squat, lunge, or similar exercise. When I assess clients, I ask them to perform a squat and tell me how it feels. If they complain of knee pain, I ask them to perform the lateral stabilization exercise with their knees bent and to then repeat the squat. Sixty percent of people no longer feel pain, a small percentage feel less pain, and a tiny proportion say they don't notice any difference (with whom I will use a different approach). This is the *wow* factor I often talk about—a simple yet extremely effective exercise. This is how we are going to build the solid foundations of our body, bearing in mind that we are only as strong as our weakest link. The key words are **mobility** and **stability**.

Let's take a look at why this exercise is so effective. We are simply activating the LL (Lateral Line) that acts as a lateral stabilizer. Its postural role is to balance the SFL/SBL. It has a movement function and acts as an adjustable "brake" for lateral movements and rotations of the trunk. That is why the position of the foot is important; the LL begins below the base of the first and fifth metatarsals (peroneal muscles).

I like to do this exercise directly on the ground (without a mat), which is why I am often referred to as "the iron lady". There is one very simple reason for this: A hard floor gives us instant feedback on our trigger points. Tension on tension. Eliminate the tension, and the pain may decrease the next time you repeat the exercise. Do you want to miss this chance to feel your body change?

Climbing the Stairs on Your Knees

Starting Position

Sit on both your heels with your feet flat, and hold your knees with your hands. If you have difficulty sitting on your heels, perform the exercise on page 95 (FEEL IT strategy): It will really help you.

Movement

Use your hand to help lift your right knee and right side of your pelvis, then raise your left knee and left side of your pelvis as you lower the right side. Continue alternating sides and increasing the ROM of your knees. Imagine you are climbing the stairs on your knees.

Focus On

- Performing fluid movements.
- Gradually increasing the level of difficulty.

Work On

- Stretching the lower part of the SFL.
- Involving the ankle, hip, and sacroiliac joint.

Discussion

I decided to include this exercise in the Functional MOBILITY strategy because, although it involves significant stretching of the lower part of the SFL, it also works multiple joints (ankle, hip, and sacroiliac joint).

Strategy
MOBILITY – Functional mobility of the hip

Starting Position
Lie face down on the ground with your hands crossed behind your back and your right leg bent. Place your right ankle on the calf muscle of your left leg. Look forward.

Movement
Slowly lift your right knee off the floor, keeping your foot and torso as still as possible. Lower your knee back to the floor and repeat as often as desired, then change sides.

Focus On
- Only lifting the bent knee: Keep your right foot resting on your left leg.
- Do not lift your pelvis; it should remain in the starting position.

Work On
- Functional mobility of the hip.
- Strengthening the hip external rotator muscles.
- Stretching the hip internal rotator muscles.

The Hip

The hip is the region that joins the pelvic area of the trunk to the lower limbs. The hip, also known as the coxofemoral or acetabulofemoral joint, is a joint with a wide range of motion (ball-and-socket joint) between the acetabulum of the iliac bone and the head of the femur.

List of Hip Internal or Medial Rotators

The key components in this action are as follows:
- Tensor fasciae latae: Internally rotates, flexes, and abducts the hip
- Gluteus minimus: Abducts, internally rotates, and flexes the thigh
- Gluteus medius (anterior fascia): Internally rotates and abducts the thigh, participating in flexion and extension
- Adductor magnus (inferior fascia, posterior by origin): Supports internal rotation

List of Hip External or Lateral Rotators

The key components in this action are as follows:
- Piriformis: Externally rotates the femur and extends the sacrum
- Superior gemellus: Externally rotates the femur
- Inferior gemellus: Externally rotates the femur
- Obturator internus: Externally rotates the femur and stabilizes the pelvis
- Obturator externus: Externally rotates the femur and stabilizes the pelvis
- Quadratus femoris: Externally rotates the femur
- Gluteus maximus: Extends and externally rotates the femur, involved in both adduction and abduction

Discussion

The position seems very comfortable, but it is the knee movement that is difficult.
Many beginners are unable to lift their knee off the ground. Try to reduce the level of difficulty (see below). Train with a change tailored for you and then try the standard exercise again.

Reduce the Level of Difficulty

- Change the position of your arms, placing them under your forehead.
- If you find it difficult to adopt the starting position, raise your pelvis slightly off the ground with a cushion. If it is still difficult, choose a different exercise and try again in a few months.

Increase the Level of Difficulty

Turn your head in the direction of your bent leg. You will see just how challenging it becomes.

Transverse Seated Windshield Wipers

Starting Position

Sit down, with your legs bent and spread slightly more than shoulder-width apart. Place your hands on the floor behind your back, and straighten your back as much as possible.

Movement

Keeping your back straight, raise your arms in front of your chest, lower both your knees down to the ground to the left, and stretch your torso upward.

Return to the starting position, but with your arms still raised, and lower your knees to the ground to the right. Continue to move from one side to the other and complete as many repetitions as desired.

Focus On

- Active kite.
- Knees on the floor.

Work On

- Functional mobility of the hip.
- Strengthening the hip external and internal rotator muscles.
- Stretching the hip external and internal rotator muscles.

Reduce the Level of Difficulty

If you are unable to keep your balance with your back straight, move your legs with your hands on the floor, supporting the stretching of the spine (see starting position).

Discussion

We are now starting to increase the difficulty level. For me, this movement is like a dance: Elegant and fluid. It is a great transitional exercise for those to come. The best way to describe it is like the cogs of a mechanical watch, where each of the different steps is perfectly coordinated. It is an important exercise for mobilizing the hip on various planes. It stretches and stimulates the internal and external rotator muscles of the hip (see box on page 124), as well as the pelvic floor. From a myofascial perspective, it works the LL/DFL (other muscles and lines are stimulated but we are focusing on the main ones).

Sideways Tilting of the Upper Body in Standing Position

Strategy
MOBILITY – Mobilization of the lumbar spine

Starting Position

Stand up straight, with your legs hip-width apart and your arms raised to shoulder height.

Movement

Imagine that you want to touch a wall to your right with your fingertips. Your right hand drags your whole upper body with it, moving your right shoulder away from the left iliac crest. Keep your arms straight and at shoulder height at all times. This sideways movement stretches the FL; use this pre-tension to elastically return to the starting position. Change sides and repeat as often as desired.

Discussion

This movement requires particular care and awareness because it is not easy to hold the stretch when moving sideways; the tendency is to bend sideways. The goal is to mobilize the lumbar region but at the same time stretch and enhance the flexibility of the area between the thoracic spine and the sacrum (the thoracolumbar fascia and the sacrolumbar fascia) where lumbar tension often builds up, thereby increasing tissue oxygenation.

Focus On

- Active kite.
- Only move your upper body. Keep your pelvis still.

Work On

- Functional mobility of the lumbar spine.
- Dynamic stretching of all the SFAL/SBL/FL, the thoracolumbar fascia, and the sacrolumbar fascia.
- Increased oxygenation and blood circulation.

Strategy
MOBILITY – Mobilization of the thoracic spine

Starting Position

Stand up straight, with your legs together. Cross your arms in front of your chest and place your hands on your shoulders.

Movement

Stretch and tilt your upper body to the right, moving your left foot behind your right leg. Then return by moving your left leg so that your feet are shoulder-width apart. Tilt your upper body to the right again, but this time move your left foot in front of your right leg. Continue these movements, focusing on tilting your upper body to the side. Tilt to the side 10 times and then switch sides.

Focus On

- Active kite.
- Stretching and then tilting.

Work On

- Mobility of the thoracic spine.
- Dynamic stretching of the Lateral Line.

Discussion
The goal of this exercise is to mobilize the thoracic vertebrae when leaning sideways, while the movements of your foot slide on a second plane. It is clear that the LL benefits from these movements by stretching. As you tilt to your right, the ribs on your left side rise toward the ceiling; take care not to tilt too far to your right. Remember: We are looking for expansion and not compression.

Reduce the Level of Difficulty

Tilt to the side without moving your legs.

Change Stimulus

Chapter 6 about fascial stretching describes an exercise called "Bamboo" (page 180), which focuses on stretching the Lateral Line.

Tired Cat on All Fours

Starting Position

Get onto all fours, with your knees perpendicular and hip-width apart, your hands shoulder-width apart and perpendicular to the shoulders, and your fingers spread. Bring your right hand and right knee together slightly.

Movement

Let your upper body and head drop between your straight arms, with your back taking on a shape like a hammock.

Move your pelvis toward the wall behind you, and keep it as parallel as possible to the floor.

Move back to the starting position, and, without stopping, move forward so that your shoulders are in front of your hands.

Focus On

- Fixed point: pelvis; mobile point: spine.
- Upper body and head dipped between your arms in "hammock" position.
- Arms straight and elbows facing outward.
- Pelvis in a neutral position. Rotating the pelvis will support the movement and the rotation of the vertebrae will be less, resulting in less mobilization. If you shorten one side (your pelvis moves toward the ribs on one side), axial stretching will be lost, and the muscles that prevent the vertebrae from rotating will be activated. Is it now obvious why it is important to keep the pelvis on its axis?

Work On

Functional mobility of the spine when unloading.

Discussion

This reminds me of the saying "small things make a big difference". The movement described above could not be summed up any better: It is small but with a great impact on the well-being of the spine.

By allowing your upper body to drop completely between your straight arms and with your pelvis in a neutral position, the vertebrae are free to rotate. This is the most difficult aspect of the exercise because we find it difficult to completely relax and let go.

We try to maintain a correct body posture (or at least a posture that seem correct to us), keeping everything under control down to the smallest detail. There is a constant demand on the muscles to keep us in a certain position, meaning that we are no longer used to "letting go", to relaxing. A balanced posture does not struggle to stay upright.

This is an exercise where we should simply let go between our arms, while the back muscles relax and allow the vertebrae to move freely.

It is particularly interesting for people with scoliosis because the movement is unloaded.

Strategy
MOBILITY – Mobilization of the spinal column

Starting Position and Movement

Get down onto all fours, push your left heel backward, and slide your left forefoot on the floor until the left leg is fully extended. Imagine: Your heel moves away from the top of your head, and the top of your head moves away from your heel (two mobile points = opposing points), creating tension (pulling) through the entire back surface of your body (SBL = Superficial Back Line). Bend your right arm and put your hand on the back of your neck (nape).

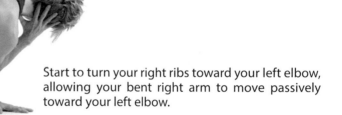

Start to turn your right ribs toward your left elbow, allowing your bent right arm to move passively toward your left elbow.

Start to move back by moving your right elbow outward; continue to turn upward, without stopping at the starting position, until the tension stops you, keeping the thoracic kyphosis and the shoulder blades far apart. Repeat the entire movement from the beginning as often as desired. Complete the exercise by turning in the opposite direction.

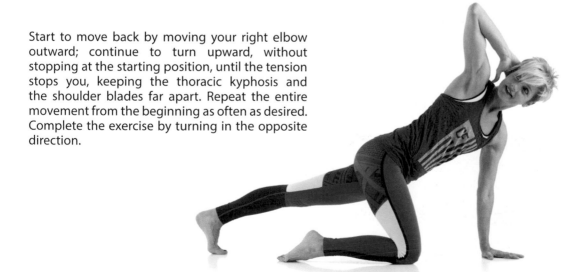

Focus On

- Active kite.
- Turning the elbow of your supporting arm outward.
- Opposing points: Heel (straight leg) and top of the head.
- Creating tension from your heel to the top of your head.
- Rotate around the central axis of your back; imagine that you are a chicken on a spit that is turning, without moving, around the rod.

Work On

- Functional mobility of the thoracic spine.
- Myofascial stretching of the LL/SL/DFL.
- Strengthening of the core, LL/SL/SBL.

Discussion

If you remember, the goal of this exercise is to mobilize the spinal column when stretched. This means no excessive muscle strain because this will block the movement. But we are also not looking for hyper-rotation, which involves excessive extension of the thoracic spine (maintain the physiological curves of the spine and rotate around them). Your gaze follows the movement of your elbow as this will also involve the cervical spine (*eye–spine connection*). This all makes for an excellent workout for the entire *core*.

Keep the thoracic kyphosis and shoulder blades far away from each other as you rotate upward; it will be slightly more difficult to rotate in this position, but this will more effectively maintain the expansion of the chest to aid breathing.

Let me explain: To create volume in the DFL, if I bring my shoulder blades together and extend my thoracic spine as I rotate upward, I reduce the space in my chest (the sternum and thoracic vertebrae come closer together); but if I maintain my thoracic kyphosis and rotate around the central axis of the spinal column, I create space in the rib cage.

Change Stimulus

As you turn your chest upward, move your gaze (just your gaze and not your head) downward and vice versa as you go back down. This stimulates the nervous system, benefitting the thoracic spine.

Reduce the Level of Difficulty

The whole exercise can be performed on all fours or in turtle position as shown below.

Spider Woman

Starting Position

Stand up straight, with your legs hip-width apart. Bend your right arm, and stretch out and spread your fingers. Create space between the hypothenar eminence and the thenar eminence.

Then extend your wrist, and tilt your head to the right.

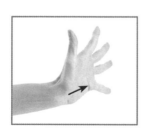

Movement

Breathe in, and move your right wrist away from your right ear and tilt your head to the left, stretching the entire arm. Keep your fingers straight until the end of the movement. If you can, extend your arm completely, with your elbow pointing backward, and bend it again; move your ear toward your wrist. Continue rhythmically for 10 repetitions and gradually increase the range of motion.

Focus On

- Opposing points: ears, wrist, and fingers.
- Keeping your wrist extended at all times, creating space in the palm of your hand.
- Fingers extended right to the fingertips and slightly spread.
- Active kite.
- Fluid and rhythmic movements, without effort or pain. The movement will only be effective if complete range of motion is achieved so as to find the desired stretching position. The extensors will then be simultaneously activated and strengthened.

Work On

- Neuro-myofascial mobilization.
- Stimulating the conscious and unconscious "conversation" between the central nervous system, tendon, and fascia muscle.
- Stretching the hand flexors and strengthening of the hand extensors.
- The goal of this exercise is to get onto all fours with your hands flat without experiencing any pain. It is also an exercise of control to check whether the imbalances have been eliminated.

Change Stimulus

By holding the position with your arm straight, the exercise becomes one of dynamic stretching. It is always better to start with mobilizations because stretching that is too violent could damage the nerve structures.

Rotate your wrist in a clockwise and counterclockwise direction, accentuating each position: ulnar deviation, flexion, radial deviation, and extension.

The extensors extend the hand and straighten the fingers. In contrast, the flexors allow the hand to close.

Discussion

The Spider Woman is a great exercise for preparing to support your weight on your hands, such as when on all fours or doing a front plank.

Through all my years of working, I have met many students who complained of wrist pain when supporting their weight on their hands. To help them, I used to place balls, boosters, or small weights under their wrists, or I gave them the option of putting all their weight on their fists, to reduce the angle between hand and forearm. Today I only use this approach in extreme cases. Instead, I start immediately with exercises to improve support, and this is just one of many that you will find in this book. If we continually use supports, muscle imbalances will never be addressed.

The functional length of muscles and the elasticity of fascial tissue are key to daily motor activity. If a joint's full range of motion is not regularly used, the extensibility of the corresponding muscles and fascial structures will decrease: They "shorten". Hypertrophic muscles, with the consequent increase in resting tension, tend to "shorten". Their antagonists tend to weaken, creating an imbalance in the joint known as postural imbalance.

The flexors of the hand are often hypertrophic and shortened. One possible cause is repetitive movement of the wrist and fingers, associated with working long hours on the computer, as well as with sports like baseball, hockey, and even tennis, not to mention many other activities in which the fingers and wrist are in a flexed position most of the time. Such activities may hinder your ability to extend your hand.

It is now clear why getting onto all fours could cause discomfort, as the ability to extend your hand is reduced. Being able to support your weight on your flat hands without experiencing pain is the goal of this exercise and a step toward eliminating the imbalance.

It is here that we very often take the wrong path. We let our limitations deceive us as we try to position ourselves more comfortably so that we don't feel any pain, often supporting ourselves on closed fists. This may be a mistake. Without realizing it, we further promote the closing and muscle imbalance of the forearm, strengthening the flexors of the hand. This increases the risk of suffering from so-called tennis elbow.

This imbalance between the flexors and extensors of the hand cannot be resolved simply by stretching the shortened flexor muscles of the wrist; the weak extensors of the wrist must also be strengthened.

To demonstrate the effectiveness of this type of exercise, I would like to tell you the story of Loredana, known as Lori, a client of mine and a close friend. Having a hairstyling studio and working as a hairdresser, she used and overworked her neck, arms, and hands. She had pain in her left thumb, which became so bad that she was barely able to hold the hair dryer. In her functional training sessions she constantly complained of thumb pain and would always try to find a comfortable, pain-free position. I then stopped all the exercises that could cause her pain and replaced them with exercises to stretch and strengthen the Arm Lines. I gave her a task to do during the day: Complete 10 to 15 repetitions of the Spider Woman exercise 2 or 3 times during her working day. The improvement was instant. This is what makes my job so rewarding: To search, create, and amend in order to make someone functional; to use science to generate changes that guarantee a *WOW* factor!

Who could benefit from the Spider Woman exercise? Everybody. Particularly anyone with carpal tunnel syndrome, neck tension, or shoulder pain.

Carpal tunnel syndrome is a very common nerve condition that is particularly prevalent in women and in certain occupations. It is characterized by pain, weakness, and numbness in the hands and fingers, which then radiates along the arms toward the shoulders. It is caused by the narrowing of the carpal tunnel due to swelling or other conditions, which compresses the median nerve inside.

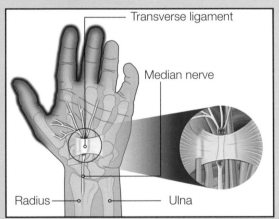

Another condition that often causes shoulder, arm, and hand pain is **thoracic outlet syndrome**. This involves the blocking of one or more of the structures that pass through the thoracic outlet: The subclavian artery or vein, the brachial plexus, or the thoracic duct. Postural imbalance of the scapulohumeral joint promotes this obstruction.

People who find this exercise interesting may also be interested in the sensory ball exercises (see page 284: Pectoral Movements Lying on Your Front).

Strategy
MOBILITY – Mobilization of the scapulohumeral joint

Starting Position and Movement

Get on all fours, with your knees hip-width apart and perpendicular to them. Place your hands shoulder-width apart and perpendicular to them. Spread your fingers.

With your arms straight, lower your chest toward the floor as your shoulder blades come together.

Raise your thoracic spine toward the ceiling with a fluid motion; you should feel your shoulder blades moving away from each other. Repeat with fluid movements as often as desired.

Focus On

- Keeping your arms straight and elbows facing outward.
- Keeping your shoulders away from your ears.
- Concentrating on the movement of your shoulder blades.
- Start with small movements, increasing the repetitions.

Wow Factor Test

Let's see how this simple exercise can improve the functional mobility of the shoulder joint (scapulohumeral joint).

Find a comfortable position and try to bring your hands together behind your back (on both sides).

Make a note of how far you were able to reach on both sides and where you felt tension.

Now perform the exercise correctly, focusing on all the points (see "Focus On") for 10 repetitions, and then repeat the test on both sides. Has anything changed? Most of my "guinea pigs" find that their mobility and flexibility has improved as the distance between their hands decreases.

Remember that it is still an improvement even if only a fraction of an inch. This will further improve each day that you work at it. It's a bit like collecting reward points at the supermarket: You have to invest to be rewarded.

Discussion

I like this exercise because it is simple, gives instant results, and also requires precision. I often see people doing many warm-up exercises for their cardiovascular system and legs, but what about the shoulders? We are then surprised that so many people suffer from shoulder pain sooner or later.

Change Stimulus

- To increase the load, perform the same exercise, but this time extend your right arm (with your hand closed in a fist) in front of your head. Repeat as often as desired, and change sides.

- Further increase the load by also extending your left leg. The position of your pelvis should not change, remaining perfectly aligned like when on all fours. Repeat as often as desired, and change sides.

- The exercise can also be performed in the plank position (front plank with legs straight: See figure at the top of page 189), but be aware that too much load reduces mobilization and only increases muscle work. This may be useful for some athletes but only if the foundations have already been laid.

- In the starting position (see first figure on page 133), draw circles with your shoulder blades. Bring your shoulder blades as close together as possible without losing your thoracic curve (kyphosis), then move them toward your ears as far as they will go. Finally, move them away from each other and then move them toward your pelvis as far as they will go. The goal is to mobilize the entire ROM of your shoulder blades in isolation without changing the position of your back.

5.4.2. MOBILITY Exercises Against the Wall

5.4.2.1. Functional Mobility: Hip

Moving Stork Against the Wall

Starting Position and Movement

Adopt the Stork position and place your hands on the wall in front of you at shoulder height. Bend your left leg by raising your knee.

Keeping your arms still, move your left knee and pelvis to the left side. Return to the starting position and turn to the right. Repeat 10 times and then change legs.

Focus On

- Fixed points: standing foot and hands; mobile point: raised knee.
- Keep your standing leg straight.
- Active kite.

Discussion

This is a fantastic warm-up exercise.

Work On

Functional mobility of the tibiotarsal joint.

One Leg Deadlift Against the Wall

Strategy
MOBILITY – Functional mobility – Hip and shoulder

Starting Position

Standing upright, place your hands on the wall shoulder-width apart and slightly below shoulder height, with your arms straight. Take two steps backward.

Movement

Push your hands against the wall, tilt your upper body forward, and, at the same time, lift your right foot off the ground by bending your right knee. As your upper body continues to tilt forward, push your pelvis backward and simultaneously extend your right leg so as to create a straight line from your toes to the top of your head. Stretch as much as you can and then return to the starting position. Complete the desired number of repetitions and then switch legs.

Focus On

- Opposing points: hands and raised foot.
- Active kite.
- Screwed humeral head (slight external rotation of the humerus).
- Knee of the standing leg slightly bent.
- Neutral position of the spine.

Work on

- Functional mobility of the hip and shoulder.
- Dynamic stretching of the SBL, the lower part of the standing leg/hamstrings/SFL/DFL/AL.
- Strengthening of the SBL/LL/FL.

Discussion

Many think that this exercise does more than just mobilize the hip. Correct. But sometimes we have to find a happy compromise: When the exercises become complicated, it is hip mobility that is likely to suffer.

The One Leg Deadlift has become a key part of my training sessions. I love it, whether using my body weight or a kettlebell. It gives me a feeling of skill, control, strength, and stretching. At the same time, I feel my glutes and hamstrings toning up nicely; it's a great exercise for the summer!

There is just one small problem... it isn't easy and requires a great deal of control and balance. The One Leg Deadlift should be attempted gradually, starting with preparatory exercises to help activate the myofascial lines.

Why choose this exercise:

- It helps with stability because you are supporting yourself with your hands on the wall.
- It teaches you the movement of the hip.
- It helps you to keep your spine in a neutral position.
- The contact of your hands with the wall helps to activate the latissimus dorsi, pushing your shoulder blades toward your buttocks.

Change Stimulus

- Take your right hand off the wall, and extend your right arm toward the floor. This should keep your right shoulder blade active (push it toward your left buttock), thereby activating the Back Functional Line (latissimus dorsi, thoracolumbar fascia, sacral fascia, gluteus maximus, vastus lateralis, patellar tendon).
- Use a resistance band (or a tube, which is a resistance band with handles), tie a knot at one end and put your right arm through it so that it wraps around your right shoulder. Grasp the resistance band with your left hand (fist) and pull it tight; place your fist on your left thigh. This will help you to visualize the path of the FL and to activate it. Perform the exercise described on page 136.

Circles With Your Arm Against the Wall

Strategy
MOBILITY – Mobilization of the scapulohumeral joint

Starting Position

Stand up straight sideways to the wall and just a few inches away from it. Stand with your feet hip-width apart and your arms straight and down by your sides. Place the back of your right hand against the wall with your fingers outstretched.

Movement

Start to make a circle around your shoulder. Continue the circle, keeping your arm straight, until you have slightly gone past your head, increasing the stretch on your pectoral muscles and on your latissimus dorsi. Now turn your hand so that your palm is touching the wall. Continue the circle.

Continue to turn until your arm is back down by your side. Turn your hand so that the back of your hand is touching the wall and start a new circle for the desired number of repetitions, then switch sides.

Focus On

- Fixed point: pelvis; mobile point: hand.
- Active kite.
- Constant and uninterrupted contact between hand and wall.
- Neutral position of your pelvis and thoracic-lumbar spine.

Work On

- Mobilizing the scapulohumeral joint.
- Stretching the Arm Lines.
- Mobilizing the thoracic spine.

Reduce the Level of Difficulty

If you find it difficult to make the circles, move away from the wall slightly and raise your arm as far as you can, then lower it and repeat, increasing the ROM slightly with each repetition.

Discussion

Caution: It may seem like a simple exercise, but the devil is in the detail. As a coach, I always have a clinical eye, and nothing ever escapes me (according to my clients). But if you want results, no compensations. Let's take a look at the most common compensations adopted:

- Raising your shoulder toward your head, giving rise to the incorrect position of the humeral head and reducing the subacromial space through which the supraspinatus tendon passes.
- Humeral head moving toward the acromion, constricting the supraspinatus tendon, thereby creating a number of problems (rubbing, inflammation, bursitis).

Did you know that, in the event of chronic inflammation of the tendons or continuous shoulder overload, the tissue turns into bone tissue?

The build-up of calcium deposits in the tendons of the rotator cuff of the shoulder causes calcific tendinitis, a condition characterized by the inflammation and calcification of the tendon in which these deposits form. This causes shoulder pain because the calcium irritates the tendons. Some authors even claim that the tendon cells undergo a process of metaplasia, transforming them into calcium-producing cells.

The calcium deposits cause the tendon to swell. When you raise your arm above your head, this swelling causes the tendon to compress below the acromion, resulting in pain.

As you raise your arm up to and beyond your head, pushing your chest forward (accentuating lumbar lordosis and the extension of the thoracic spine) is a way to compensate because it brings the origins and insertions of the latissimus dorsi, pectoralis major, biceps brachii, and so on closer together. That is why it is important to breathe into the thoracic-lumbar region (T12). Go against your natural tendency. Expand your back away from the sternum.

Choose the exercise that is most suitable for your ability, commit 100 percent, and you will be rewarded and amazed at how quickly you see and feel the results.

Change Stimulus

- Keep your upper body still at the beginning so that the focus of the movement is on joint mobility of the scapulohumeral joint.
- Use your thoracic spine as you start to turn your upper body with your arm and head to also provide functional mobility of your thoracic spine.
- Take one step forward with your left foot, keeping both your heels on the floor. Push your right foot into the floor as you raise your arm. As well as joint mobility, you should feel a net increase in tension in your rectus abdominis and hip flexors (SFL/DFL).
- Take one step forward with your right foot, keeping both your heels on the floor. Push your left foot gently into the floor as you raise your arm. Everything changes! You should really feel the FL from your left inner thigh to your right pectoralis major.
- Screwdriver to the wall: Stand up straight sideways to the wall and about 1.5 feet (half a meter) away from it, feet hip-width apart, arms straight and down by your side. Start the circle with the back of your hand and your nails against the wall. Slide your hand up the wall until you reach the height of your ears. Start the circle with the back of your hand and your nails against the wall. Slide your hand up the wall until your arm is parallel to the floor. Now externally rotate your arm so that your thumb is against the wall as you continue to move your arm up the wall. When your fingers are pointing to the sky, internally rotate your arm, and place the palm of your hand on the wall. Continue moving your arm along the wall until your arm is back down by your side. Maintain an active kite throughout the movement.

Bridge Against the Wall

Strategy
MOBILITY – Mobilization of the spinal column and much more

Starting Position

Lie down on your back, and place your feet against the wall, hip-width apart. Keep your arms straight and by your sides and your elbows facing outward. Keep your spine in a neutral position.

Focus On

- Fixed points: hands and feet; mobile points: knees and back.
- Push your arms downward as you descend; this will help your back to sink gradually toward the floor.

When Going Up, Work On

- Strengthening the extensor chain of the hip and arms.
- Dynamic stretching of the SFL and AL.

When Coming Down, Work On

- Functional mobility of the spinal column.
- Dynamic stretching of a part of the SBL.
- Elimination of tension in the thoracic-lumbar spine.
- Improving breathing.

Discussion
An interesting exercise that will help you to understand how forces are distributed. If you push with your legs, you will only work these and you will move away from the wall. Balance the force between your legs and your arms.

Movement

Breathe out. With an explosive but controlled movement, lift your pelvis up and push your hands against the floor and your knees forward. Keep the upper edge of your shoulder blades on the floor. You should feel your hamstrings and glutes activating. Keep your arms active, relax your glutes, and let your spine sink between your arms toward the floor to return to the starting position. Repeat as often as desired.

Change Stimulus

When coming down, move your upper body slightly to the right and come down one vertebra, then move to the left side and come down another vertebra. Repeat until your pelvis is on the floor.

Increase the Level of Difficulty

Raise your left leg and stretch it toward the ceiling. Then repeat. Switch legs. Now all that's left is to put both legs into the air at the same time... Only joking! (At least, I would never do it.)

5.4.3. MOBILITY Exercises With Equipment

5.4.3.1 Functional Mobility: Hip

Circles With the Resistance Band

> **Strategy**
> MOBILITY – Functional mobility of the hip

Starting Position

Lie on your back, with your legs hip-width apart. Slide your right foot into the loop of the resistance band. Hold the band with your hands, with your left hand grasping the end of the band. Move your arms toward your head, keeping them within your field of vision. Keep your back in a neutral position. Keep your arms shoulder-width apart.

Discussion

I first saw this exercise performed by Gray Cook in 2009 in the USA, and it remains a great way to mobilize the hip and stabilize the trunk, avoiding the tendency to tilt to one side. We can certainly say that it mobilizes the hip but that at the same time it offers lateral stabilization of the scapulohumeral joint. And let's not forget the arms, which are almost in an overhead position.

Let me add a side note here: As you turn your right leg outward, actively anchor your left hip to the floor in opposition. This will increase the stretch and therefore improve mobility.

This exercise, which is suitable for most people, is a great warm-up exercise and can improve squats, as well as increase the ROM and prevent osteoarthritis of the hip.

Change Stimulus

Use a resistance band with greater resistance or a tube (sliding your foot into one of the handles).

Movement

Keep your right leg straight, and lift it up.

Bend your right leg by moving your knee outward, place the outer edge of your foot on the ground close to your left thigh, and begin to slide your foot forward until your leg is fully extended. Repeat 5 times, and then change direction. Perform the same movement in the opposite direction another 5 times. Finally, repeat the entire exercise with the other leg.

Focus On

*   Fixed point: pelvis; mobile point: leg.
*   Do not move your arms throughout the entire exercise.
*   Keep your back in a neutral position (T12 firmly on the floor).

Work On

*   Functional mobility of the hip.
*   Strengthening the muscles of the arms, legs, and core.
*   Myofascial stretching: SBL/DFL.
*   Coordination.

Aircraft Marshalling With Clubs

Strategy

MOBILITY - Functional mobility - Scapulohumeral joint

Starting Position

Stand up straight with your legs hip-width apart. Hold the thin end of the clubs and cross them in front of your chest: Right arm in front and left arm behind.

Movement

In one smooth motion, also cross your arms in front of your chest and bring your elbows close together. Continue the circular movement. Uncross your arms and bring your hands to the back of your neck, raising your elbows as you do so.

Open your elbows as wide as possible: You should feel your chest stretching. Keep your ribs closed and your shoulder blades active.

Now stretch your arms upward toward the corners of the room.

Lower your arms down by your side, and repeat the exercise, changing which club is in front when you cross them each time. Repeat the entire sequence 10 times.

Focus On

- Active kite.
- Fluid, uninterrupted movements.
- Slow upward movements and quick downward movements.

Work On

- Mobilizing the scapulohumeral joint.
- Stretching and elasticity of the arms and trunk.
- Coordination.

Discussion

As the name suggests, the exercise is inspired by aircraft marshalling signals, which have been internationally standardized by the International Civil Aviation Organization (ICAO).

It is a great exercise for warming up the scapulohumeral joint and improving mobility. Bear in mind that you will need to train hard to master the movements and make them fluid.

Increase the Level of Difficulty

Use heavier clubs.

STRETCH: FASCIA STRETCHING AND SHAPE

6

STRETCH

6.1. STRETCH STRATEGY

As in the previous chapter, we will discuss in detail why and how I decided to include the Fascia Stretching strategy into my training program.

It seems that the question of how, how much, and when to stretch has finally been answered. Even if you regularly stretch your muscles, you are probably unaware (unless you are an expert) that everything has already been said about Bob Anderson's theory: Is static stretching okay, or is dynamic stretching better? Stretching before or after physical exercise?

In reality, the question to answer is much simpler: When and how much to stretch in terms of the moment in which the stretch is performed and the sport to be played?

The term stretching describes the method that is used to improve muscle-tendon-fascia flexibility (and therefore joint mobility) by means of simple or complicated stretching exercises.

As part of physical preparation, the advent of stretching exercises represented a step forward as athletes began to pay more attention to the different stresses the various muscle groups are subjected to, as well as to joint mobility. The practice of stretching has been recommended for many years and has been performed indiscriminately because it was believed that it could only be beneficial to the body. But the reality is that, like all types of

143

training, stretching is multifaceted and cannot be used arbitrarily. It is important to consider the goal that you want to achieve, when the exercise is to be performed, the sport (and therefore motor) biography of the person or athlete, the specific sport the person plays, and correct technical execution of the stretch. So get ready to join me on this eye-opening journey; I will revolutionize the way you see stretching.

The stretching exercises that I propose in this book are varied, diverse, and fun and involve the whole body in order to stimulate the entire myofascial system.

What are the benefits of stretching?

- It will improve your daily life.
- It will increase your athletic performance.
- It will preserve or improve your flexibility.
- It will help to prevent injuries.

6.2. MOVEMENT AND SHAPE

What does fascia stretching do? It moves fascial structures and gives shape to the muscle chains and organs involved. It also builds new tissue by stimulating collagen synthesis.

As stated in section 1.1.2, "Defining Fascia", the goal of the FReE method is to involve as much connective tissue and as many types of it as possible. That is why I have chosen a range of static, dynamic, dynamic active, and isometric training exercises.

Definition of Fascia Stretching

Fascia stretching is the slow or fast stretching of all or part of a myofascial line, which primarily involves the various fascial tissues and a series of other connected structures.

Why Is Fascia Stretching Important?

It hydrates the fascial tissue. This is vital because it enables the muscles, the internal organs, and the other tissues enveloped in the fascial tissue to glide freely over one another, guaranteeing the optimal transmission of load and tension information.

When the fascial tissue starts to dehydrate, it becomes tacky, sticky, or matted. When this happens, internal organs, muscles, and other tissues lose the ability to communicate with each other, which hinders the distribution of load and tension. All this creates compensations that I call "works in progress". These individual instances of stiffness can turn into patterns of tension that, over time, can give rise to symptoms such as back pain, shoulder pain, and so on.

For every movement that our body makes, our skin and fascial tissue should be free to slide over the muscles, and the muscles should be able to move from one to the other. If this is hindered, we are more susceptible to injury.

Let me give you an example. If you have not done much exercise and have been physically inactive and sedentary for a long time, we now know that your tissue will have become sticky creating stagnation. But now your niece wants to play ball with you, or your nephew wants to do yoga, and of course you can't say no. During the activity, you suffer a sprain or a tear. This happens when the fascia is dehydrated, caused by a lack of movement. Dehydration of the fascial tissue alters the fluidity of circulation of this dynamic system, disrupting the delicate balance between the rigidity and elasticity of the fascia. This also affects the sensory functions of the nerves and ultimately prevents the autonomic nervous system from supporting us, protecting us, and stabilizing us without significant strain.

When fascia is tacky, sticky, or matted (dehydrated and retracted) due to a lack of movement, a combination of static and dynamic stretching exercises will benefit the fascia enormously by making it elastic and slick or, in other words, FReE.

When I use the terms "sticky", "matted", and "tacky", I am referring to several overlapping layers of tissue that stick together, like a cashmere sweater that becomes bobbly and stiff after it has been washed in hot water. In the words of Andrew Taylor Still, the founder of osteopathy, "When the fluids of the body are stopped in the fascia, organs and other parts of the system, stagnation, fermentation, heat and general confusion will follow" [69].

What Are the Benefits of Stretching?
- It hydrates the fascial tissue.
- It rapidly improves flexibility.
- It improves posture.
- It frees and cleans the myofascial lines from tension.
- It increases the resilience of the fascial structures.
- It makes the fascia elastic and free to glide between the various tissues.
- It prevents injury.
- It helps to generate new tissue by stimulating fibroblasts to synthesize collagen (see chapter 1).

6.3. STRETCHING TECHNIQUES

If you do an Internet search you will find a range of different stretching techniques:

- From more simple to more difficult
- To be performed before, during, or after training
- Grouped according to types of people or athletes
- To be performed ideally with a coach or therapist

There are two main types of stretching:

1) **Static stretching** means holding a muscle (or muscle group) in a stretched position to maintain a maximum stretch. As the name suggests, there is no movement, and the position should be adopted as slowly as possible.

2) **Dynamic stretching** refers to controlled movements of the arms and legs that gradually take your body to the limits of its range of motion.

We will now look in detail at the four main types of stretching used in the FReE method from a myofascial perspective: static, dynamic, active, and isometric.

6.3.1. FReE - Static Stretching

The static stretching technique is based on reaching and maintaining the maximum stretching position of all or part of a myofascial line for a certain period of time.

It differs from Bob Anderson's analytical stretching in that we do not stretch individual muscles but all or part of the myofascial lines.

Until now, exercises that stretch individual muscles in isolation and in one direction have been the norm. We meticulously analyzed every single muscle, moving their origin and insertion as far away as possible from each other. Today, we know that stretching does not mean simply stretching an individual muscle and its part of the fascial tissue. Rather, stretching acts on a large net, which crosses the whole body and covers it like a wetsuit.

Current stretching techniques are based on muscle chains, or rather myofascial lines, that run from one end of the body to the other, known as end to end. Static stretching can be further classified as

- passive static stretching, and
- active static stretching.

Passive static stretching uses an external force like a wall, a resistance band, a towel, or a therapist to hold the stretch in position. No muscle group is statically contracted to hold the limb in place (see the Splits Against the Wall exercise on page 220).

Active static stretching requires the strength of antagonist muscle groups to hold the limb in place for the stretch. In this case, to move the right leg, the hamstrings and gluteus maximus are active (see the Supine Corkscrew in Action exercise on page 198).

Passive and Active Static Stretching Technique

1) Slowly adopt the full stretching position (just beyond your normal range) of all or part of a myofascial line, using the end to end strategy.
2) Continue to breathe and hold this position for between 10 and 30 seconds.
3) When the stimulus of the stretch weakens, breathe out, find a new stretching position with a new stimulus, and hold the new position.
4) Repeat 3 times.

6.3.1.1. Duration of the Stretch

How long to hold a stretch for is a topic that has been debated at length. However, it is now known that it is not so much the duration but more the continuity and intensity of the stretch that influences and improves flexibility.

Increased flexibility is evident after just 10 to 15 seconds. Increased flexibility continues up to 45 seconds, after which the curve flattens as time increases. The ratio between efficacy and time deteriorates [62].

Increasing the duration of a stretch to 2 minutes also increases the relaxation and deformation (crimping) of the tissue. The stretching effect lessens after 2 to 3 minutes, and after 2 hours the tissue will have returned to its initial level.

The most important parameters are the intensity and duration of the stretch.

6.3.1.2. Advantages and Disadvantages of Static Stretching

Advantages of Static Stretching
- It is safe and easy to learn and do.
- Energy expenditure is minimal.
- It is more relaxing.
- It can prevent problems associated with the stretch reflex.
- If done at an appropriate intensity, it can give rise to reflex muscle relaxation induced by the action of GTOs (Golgi tendon organs).
- It facilitates semi permanent structural changes in terms of elongation.
- It lowers blood pressure and heart rate.
- It acts on all or part of the myofascial lines.
- It stimulates multiple fascial components of the muscle, primarily extramuscular and parallel tissue.

Static stretching does not just improve muscle and tendon lengthening but also the range of motion of the joint. Slow and prolonged stretching has an influence on physiological changes like blood pressure and heart rate, both of which fall, inducing relaxation.

When fascial tissue is stretched, it sends signals to the autonomic nervous system, which activates the parasympathetic nervous system, which in turn induces a relaxation reaction.

Disadvantages of Static Stretching From an Agonist and Postural Perspective
- It lacks specificity.
- It does not improve coordination.
- It does not activate muscle spindles (speed sensitivity).
- It is not ideal prior to a race.

The main disadvantage of static stretching is its lack of specificity. In practice, most sports involve dynamic ballistic movements, during which the MTU (muscle-tendon unit) must support violent and repetitive stretches.

As well as being non-specific, it does not improve coordination and does not activate the primary muscle spindle endings, which are sensitive to the speed of movement.

It must be remembered that muscles possess two types of receptors: The first measures both the speed and the length of the stretch, while the second is only sensitive to changes of length. This is another reason to choose a combination of static and dynamic stretching exercises.

When Should Static Stretching Exercises Be Performed?

The bulk of the studies on static stretching before physical activity or a race suggest that it should be avoided. Numerous scientific studies on static stretching before a race have found poorer performance in terms of

- production of muscle strength,
- power,
- balance, and
- reaction time.

The contracting ability of a muscle subjected to an excessive stretching load decreases because of a change in the muscle's capacity to absorb and distribute the shock resulting from an imposed external load. This is accompanied by a change in the stiffness with which the muscle-tendon system reacts to an imposed load.

From the studies analyzed, it can be concluded that holding a stretch for 1 minute or longer immediately prior to physical activity reduces performance by between 5 and 7.5 percent. Research by the University of Wuppertal found that athletic performance fell by between 2 and 23 percent [19].

All of the above concerns static stretching prior to training. This does not mean that you should not do deep and extensive stretching. An extensive stretching session is always useful, but it should not be performed immediately before physical activity. Think of it as a training session in itself, perhaps to be performed on your rest days (see related studies from page 28 onwards chapter 1).

6.3.2. FReE – Dynamic Stretching

This type of stretching, which has long been practiced by dancers and athletes alike, was shelved because it was believed to cause injuries. This all happened in the 1980s as interest in static stretching was growing. Static stretching was believed to warm up the muscles and prevent injuries, theories that were never proven. Subsequently, a series of studies refuted these theories, and dynamic stretching as an exercise is now an integral part of physical education.

Dynamic stretching refers to the controlled movements of the arms and legs that gradually take your body to the limits of its range of motion (unlike ballistic stretching, which tends to force one part of your body beyond its range of motion). A classic example of dynamic stretching is controlled movements of the arms and legs or twisting of the trunk. To better illustrate this, I would like to give you a practical example of this type of stretch, using the Elephant exercise.

1) Adopt the starting position, moving the three opposing points away from each other: Move your right heel away from the top of your head and raise your lumbar spine, pulling the myofascial Superficial Back Line, end to end.

2) In this position, the holding resistance of the muscles involved in the stretching tension increases. The muscle contracts and, in so doing, sends an impulse to the arrangement of its internal transverse fascia. This involves more fascial tissue.

3) Breathe out; start moving your upper body, touching the inside of your right foot with your fingertips and trying to touch the floor. This increases the tension on the Superficial Back Line and stretches you to the limit of your range of motion.

4) Breathe in; bend and raise your lumbar spine.

5) Breathe out; move your fingers to the outside of your right foot.

6) Continue rhythmically 5 to 10 times.

Dynamic stretching is often recommended in sports programs that involve high-velocity movements because it acts on the elasticity of the tendons and muscles. When an agonist muscle contracts rather quickly, it tends to stretch the antagonist muscle. This type of stretching involves movements by which joint speed and range of motion gradually increase. The goal is to warm up one or multiple myofascial lines and to mobilize the joints through which these lines pass, as well as to improve dynamic flexibility, which is why dynamic stretching is particularly suited to warm-up exercises prior to training.

However, it is important to stress that in order to get the most out of a program focused on flexibility, the proposed exercises must be speed specific. This means that the stretching speed adopted in the stretching program must be as similar as possible to the speed that shall be required in performing the specific movements of the sport or discipline in question.

6.3.2.1. Advantages and Disadvantages of Dynamic Stretching

Advantages of Dynamic Stretching
- It rapidly improves flexibility.
- It hydrates the fascial tissue.
- It improves and increases the range of motion of the joints.
- It raises the body temperature and therefore the speed at which nerve impulses are conducted.
- It accelerates energy production and essentially prepares the body for physical activity.
- It acts on one or multiple myofascial lines.
- It stimulates multiple fascial components of the muscle, primarily serial, extramuscular, and parallel tissue.

In practice, while static stretching is like yoga (it tells the body to relax), dynamic stretching rouses the body, preparing it for physical activity.

It may be even more beneficial to also include some motor control and proprioception exercises, which emphasize the effects of dynamic stretching performed prior to physical activity.

I recommend using dynamic stretching after muscle tears or strains have healed, or even on scar tissue, where the fascial tissue tends to create adhesions. The goal of dynamic stretching is to rehydrate the tissue, making it slick and elastic once again. To prevent injury, it is important to start with moderate movements before gradually increasing them.

Disadvantages of Dynamic Stretching
- If performed incorrectly, it is of little use.
- Tissue may be damaged if the stretch is excessively forced.

6.3.3. FReE – Active Dynamic Stretching

Active dynamic stretching is a muscle stretching and fascial release technique that provides easy and effective stretching of the main myofascial lines. It is very interesting because it ensures the functional and physiological reactivation of superficial and deep fascial planes. This stretching technique stimulates the brain and the body to seek out new forms of movement, promoting rapid improvements in flexibility.

Examples of Active Dynamic Stretching
- Rotation of the arms
- Kicking
- Movement on different planes
- Squats
- Lunges

Let me give you a practical example. The Shift Squat exercise (bottom right of page 182) stretches the LL and the hip flexors, improving the extension of the joint and reducing the stiffness of the surrounding muscle.

Here's a simpler example that you can try now:

1) Stand up straight with your legs hip-width apart.
2) Stretch your arms out in front of you with your hands together.
3) Slide your pelvis to the left and slightly bend your legs. Hold the position for 1 to 2 seconds.
4) Return to the starting position and repeat on the other side.
5) Repeat 5 to 10 times per side.

Breathe normally or breathe in as you move your pelvis to the side.

Technique: Stretching a myofascial line for no more than 2 seconds gives your muscles an ideal stretch without triggering the protective stretch reflex and the subsequent reciprocal contraction of the antagonist muscle. This technique offers significant benefits and can be performed without excessive tension or subsequent trauma.

Recent scientific research seems to suggest that dynamic stretching prior to competitive sports should be preferred over static stretching to reduce muscle stiffness and prevent injury. This may be particularly true for athletes who partake in strength and power sports.

6.3.3.1. Advantages and Disadvantages of Active Dynamic Stretching

Advantages of Active Dynamic Stretching
- They are quick to do and have a rapid impact on flexibility.
- They are great for warming up.
- They do not inhibit strength or power activities prior to training.
- They stimulate multiple fascial components of the muscle, primarily serial, extramuscular, parallel, and transverse tissue.

Disadvantages of Active Dynamic Stretching
- They are less effective at improving flexibility and their effects do not last long.
- They are not relaxing.

6.3.4. FReE – Isometric Stretching

Isometric stretching is considered to be one of the best and most effective techniques for developing static-passive flexibility. It is widely used in dancing, martial arts, and artistic gymnastics to fully loosen the muscles. In addition, this type of technique significantly helps to reduce pain experienced when stretching.

This technique was developed by the American neurophysiologist Herman Kabat at the end of the 1940s for neuromuscular reeducation and rehabilitation and was subsequently adopted and adapted as a training exercise. This technique is also known as Contract Relax (CR) and Post-Isometric Relaxation (PIR).

There are various ways to use isometric stretching:

- PNF (Proprioceptive Neuromuscular Facilitation) is a combination of passive and isometric stretching. The strategy behind this technique consists of passively stretching a muscle group before contracting it isometrically against an immovable resistance for 7 to 15 seconds and, finally, stretching it passively again for a further 20 seconds.
- CRAC (Contract-Relax-Antagonist-Contract) involves an initial phase of passive stretching, followed by two isometric contractions, the first by the antagonist and the second by the agonist, followed by a final phase of passive stretching.
- CRS (Contract-Relax-Stretching) is a stretching technique that involves the isometric contraction of a muscle for about 10 to 15 seconds, followed by relaxation for 5 to 6 seconds, followed by stretching.

It is essential not to overdo the stretches or contractions; in other words, never force the stretch or contraction of a muscle beyond typical muscle discomfort.

In gymnastics (artistic and rhythmic) and in dancing, which both require high muscle extensibility, PNF is not necessary because the dynamic and static exercises of these disciplines provide enough stress and strain, even at a young age (a phase that is particularly sensitive to increased extensibility). Furthermore, PNF is not recommended in children under the age of 16 years.

In FReE, I use the techniques that I have described, but I modified them slightly. The difference with regard to the isometric technique is that I include it in both static and dynamic stretching. Let me explain.

Isometric Static Stretching
Adopt a position whereby all or part of a myofascial line is fully stretched, isometric contraction of the agonist of up to 20 percent of its strength for 3 to 5 seconds, followed by relaxation for 10 to 15 seconds. Repeat 3 times. You could also perform this exercise with activation of the antagonist.

Isometric Dynamic Static Stretching
Adopt a position whereby all or part of a myofascial line is fully stretched, isometric contraction of the agonist of up to 20 percent of its strength for 3 to 5 seconds, followed by active relaxation swinging for 10 to 15 seconds and increasing the range of motion. Repeat 3 times. You could also perform this exercise with activation of the antagonist.

Isometric Dynamic Stretching
Adopt a position whereby all or part of a myofascial line is fully stretched, isometric contraction of the agonist of up to 20 percent of its strength; at the same time perform the movements 5 to 10 times then relax for 10 seconds. Repeat 3 times. You could also perform this exercise with activation of the antagonist.

The potential risks of this type of stretching should not be underestimated and should in any case only be attempted by athletes who have a significant degree of myotatic (stretch) reflex sensitivity and control, such as gymnasts and dancers.

I would like to give you a practical example of isometric dynamic stretching, using the Elephant exercise (page 155).

Agonist activation:

1) Adopt the starting position, moving the three opposing points away from each other: Move your right heel away from the top of your head and raise your lumbar spine. Keep the Superficial Back Line in traction – end to end – stretched as much as possible.
2) In this position, the holding resistance of the muscles involved in the stretching tension increases. The muscle contracts and, in so doing, sends an impulse to the arrangement of its internal transverse fascia. This involves more fascial tissue.
3) With traction, lightly press your right heel into the floor (from 0 to 100 percent, use 5 to 20 percent of your strength) to increase the isometric work on the SBL, particularly the calf muscle and hamstring.
4) Then keep the muscle stretched and in tension as you sway your upper body side to side 5 to 10 times. Continue to breathe.
5) Finally, breathe out and relax the tension to increase the stretch.
6) Repeat 3 times.

Antagonist activation:
Repeat the entire exercise with your right foot raised, pointing your toes toward your tibia.

Just one exercise is sufficient for each myofascial line.

In addition, to activate the Golgi receptors and induce relaxation of the desired tissue after contraction, just 5 to 20 percent of your maximum strength is required.

This exercise may be included in a context of static stretching (holding the position and creating tension) or in dynamic stretching (as with the Elephant exercise that includes traction with the heel).

Both the isometric technique as well as increased tension in the final position of a stretch represent an interesting traction stimulus for the transverse fascia within the muscles (see page 29).

6.3.4.1. Advantages and Disadvantages of Isometric Stretching

Advantages of Isometric Stretching
- It rapidly improves flexibility.
- It stimulates multiple fascial components of the muscle, primarily serial, extramuscular, parallel, and transverse tissue (see page 29).
- It reduces pain associated with stretching.

Disadvantages of Isometric Stretching

- By excessively forcing the stretch, the integrity of the muscle-tendon-fascia is put at risk, damaging the tissue.

If you ever have the chance, watch a cat waking up: It stretches and instinctively uses stress in traction to activate the muscles and fascia. It is active stretching with muscle contraction.

Stretching and yawning is the best thing we can do when we wake up, during the day, or before physical activity because it increases blood pressure and blood oxygen levels to give us a boost.

6.4. CHARACTERISTICS OF FASCIA STRETCHING

The effect of stretching on stressed or stimulated structures has a formative stimulus: It forms tissue.

We now know that, to build a high-quality fascial network, it is not enough just to stimulate it by applying tension, but multidirectional stimuli must also be used to amplify its qualities.

Myofascial stretching exercises are characterized by

- long myofascial lines (by Thomas Myers), end to end,
- three dimensional or multiplane movements, angle variations,
- swinging movements,
- static movements,
- dynamic movements, and
- isometric work.

Three-Dimensional Movements

Fascia should be pulled and stretched in all directions, at different rhythms, and at different speeds.

Wiggling, swaying, swinging, and rocking: All these actions help to stimulate the fascia.

Let me explain what I mean by "long myofascial lines". Imagine a chain made up of multiple links (one of our myofascial lines) in which one of these links is obstructed or shortened. If I go to stretch the entire chain, the person will surely try to compensate by using the more flexible links, bypassing the shortened link.

That is why it is important to concentrate when performing the exercises and to follow these rules:

1) Listen to your body and note where it is working and where it is not working.
2) Perform the exercise correctly, using the points detailed in "Focus On", and do not lose sight of them.

6.5. STRETCHING INTENSITY AND BREATHING

By myofascial stretching intensity I mean working within the maximum stretching capability of the myofascial line. Stimulate but never work in pain.

You don't have to exceed your limits or try to break a new record every day. With stretching, we often try to quickly achieve, and even exceed, the maximum extension of a muscle or joint. This is not correct and could be damaging.

The goal is to reduce myofascial tension and promote freedom of movement, rather than to achieve maximum flexibility of movement, which often leads to over-stretching and injury. The fascia has to adapt to the new demand, and this requires patience.

Don't forget to breathe; it is an important component of every movement and helps us to improve our flexibility. Think of your chest as the center of vitality.

6.6. STRETCH STRATEGY PLAN

It is now time to get to the crux of the matter: planning a training session or appropriate inclusion of stretching exercises. We will see when to stretch in terms of the moment in which the stretch is performed and the sport to be played.

We have looked at why stretching is important, and we have seen the different stretching methods and the pros and cons of the various techniques. We will now put the pieces of the puzzle together and see how to correctly assemble them in a training session. There are many different ways to complete the puzzle, but I am going to highlight the few I consider to be the most important.

Please remember that in this chapter I am only talking about the myofascial stretching strategy and how to include it more effectively in a training context. In an actual training session I do of course include other FReE strategies, which you can find in chapter 9.

STRETCH Exercises in a FReE Training Session

Fascial stretching exercises can be found in the following phases of a FReE training session:

- Warm-up
- Core part of the training session
- End of the training session

Repetitions: 3 to 10 per exercise
Training session duration: 30 to 55 seconds

STRETCH Exercises in a General Fitness or Sport-Specific Training Session

If necessary, it may be useful to include a dynamic stretching exercise before a specific exercise. Example: Before doing pull-ups, include an exercise such as the one on page 225 to improve your performance.

STRETCH Exercises at the End of a Training Session

These exercises are useful for muscle regeneration, particularly in intense sports involving heavy loads, where the concentration of lactic acid after training is high. Oxidation in the muscle is undoubtedly the most important mechanism for eliminating lactate, which is why active recovery can increase the lactate elimination rate. Active recovery (maintaining low intensity dynamic muscle activity) has a positive effect on the cardiovascular and lymphatic systems.

Static stretching is going to compress the blood and lymphatic vessels, thereby hindering recovery. Low intensity dynamic and active dynamic stretching may be useful, but active recovery like a gentle stroll is preferable.

STRETCH Exercises in a Stretching Session

Warm-Up
Dynamic Stretching – 5 to 10 Repetitions
Isometric Dynamic Stretching – 3 Repetitions
Active Dynamic Stretching – 5 Repetitions
Passive Isometric Static Stretching – 3 Repetitions
Passive Static Stretching – 3 Repetitions
At the beginning of the session, hold the position for 10 to 30 seconds.
Repetitions are per exercise.
Session duration: 30 minutes
Recovery: 36 to 48 hours

STRETCH Exercises in a Relaxing Body and Mind Stretching Session

Warm-up
Active Static Stretching – 3 Repetitions
Passive Static Stretching – 3 Repetitions
At the beginning of the session, hold the position for 10 to 30 seconds.
From the middle of the session, hold for 10 to 30 seconds or up to 1 minute.
Session duration: 55 minutes
Recovery: 36 to 48 hours

In the end variety is the key to hydrated and young fascia.

Static and Dynamic Stretching

Techniques	Seconds	Repetitions
Passive static stretching	10-30 sec	3
Active static stretching	10-30 sec	3
Prolonged static stretching	1 min	1
Dynamic stretching	1 sec	5-10
Active dynamic stretching	1-2 sec	5-10
Isometric stretching	3-5 sec	3

6.7. THE INDIVIDUAL AT THE HEART OF OUR CHOICES

Let's now try to reflect by comparing different athletes and people. Let's compare two athletes, with significantly different structures. If I am helping to train a resistance athlete who has a high proportion of red fibers (high oxygen content), I would recommend a long, high intensity stretching session but only once or twice a week.

Why? The muscles are full of oxygen, meaning that they will fatigue more slowly. That is why the intensity and duration can be higher than for a strength athlete. If we increase the intensity, we reduce the frequency.

For a strength athlete with a high proportion of white fibers, I would recommend the following:

- Intensity: less intense
- Duration: not as long
- Weekly frequency: higher

Why? The muscles contain little oxygen and will tire quickly. That is why the intensity and duration will be lower, but the frequency will be higher to give a training stimulus.

This also applies to less flexible people; I prefer to work with them as I do with strength athletes.

6.8. FASCIA STRETCHING EXERCISES

We will now take a detailed look at fascia stretching exercises, which differ from each other in terms of myofascial lines involved, starting position, and equipment used.

Myofascial lines involved: SBL - SFL - LL - SL - DFL - AL
Starting positions: Standing up straight – half kneeling – on all fours – sitting – lying on your back – lying on your front
Equipment used: Body weight – wall – equipment

For all myofascial sliding movements, think of your skin like a wetsuit that covers your entire body, from your fingers and toes to your head.
When you perform swinging movements like in the Elephant exercise, imagine that you are slipping, sliding, and expanding into this wetsuit.

6.8.1. Fascia Stretching of the SBL

Elephant

Strategy

Dynamic stretching of the SBL

Starting Position

Stand up straight, with your feet hip-width apart. Take one step forward with your right leg, keeping your left leg straight and your left foot firmly on the floor. Lean forward with your upper body and try to touch the floor with your fingertips. Perform a maximum stretch.

Movement

Breathe out: Start moving your upper body, touching the inside of your right foot with your fingertips and trying to touch the floor. Breathe in: Bend and raise your lumbar spine (see figure on the right) and breathe out as you move your fingers to the outside of your right foot. Repeat as often as desired. Repeat the entire exercise with your left foot forward.

Focus On

*   Expanding within your skin.
*   Keeping both legs straight.
*   Moving your upper body, which will cause your arms to move.
*   Bending your lumbar spine.
*   Letting your head drop toward the floor.
*   Keeping the pelvis level.

Work On

Primarily the flexibility and elasticity of the entire myofascial Superficial Back Line (SBL).

Change Stimulus

Isometric dynamic stretching: Apply gentle pressure and traction with your right heel. Continue moving your upper body.

Discussion

This exercise seems quite simple, but if you are not flexible enough (at the moment) you will not be able to touch the floor. Try to improve: Give your fascial network time to adapt to new demands. If you make a sudden and excessively intense movement, the fascia reacts, but in the opposite way (i.e., it stiffens), hindering your efforts to make it slick and elastic. After a few repetitions you should feel the movement becoming more fluid, and more elastic, and you should be getting closer to the floor. If you feel pain in your lumbar spine, stop and choose a different exercise because this is a sign that you are not yet ready.

Round Elephant

Strategy
Dynamic stretching of the SBL

Starting Position
Take one step forward with your right leg and point your right foot upward, keeping your heel on the floor. Keep your left leg straight and your left foot firmly on the floor. Lean forward with your upper body, letting your left leg bend if needed, and try to touch the toes of your right foot. Perform a maximum stretch.

Movement
Start moving your upper body and try to draw a circle around the toes of your right foot with your fingertips. With a fluid and circular motion, move your upper body in circles around your toes. Complete 5 to 10 circles in a clockwise and counterclockwise direction.

Discussion
This exercise is a variation of the previous exercise. Changing the position of your front foot increases the stretch on the SBL. The circular movements change the stimulus.

If you lift your hips (pelvic bone) upward as you perform the exercise, you will increase the stretch on your hamstrings (back of your thigh) but reduce the stretch on your back. The goal is not really to touch the floor at all costs but rather to expand the entire myofascial line, which in this case is the SBL. It is important to remember that the DFL is also part of the lumbar spine and passes through it (see page 47 onward). And where do more compressions occur?

Focus On
- Expanding within your skin.
- Keeping both legs straight.
- Starting the movement from your upper body.
- Bending your lumbar spine.
- Letting your head drop toward the floor.
- Keeping the pelvis stable.

Change Stimulus
Speaking of isometric training (see page 149 and onward), I have described the various techniques of this exercise in detail.

Strategy

Dynamic fascia stretching of the SBL

Starting Position

Stand up straight, with your feet hip-width apart. Take one step forward with your right leg, keeping your left leg straight and your left foot firmly on the floor. Lean forward and downward with your upper body, touching the floor with your fingertips. Let your head drop toward the floor, pulling the entire SBL, end to end.

Movement

Start the movement by pushing your right knee forward (in the direction of your toes), going as far past your toes as possible. To return to the starting position, start the movement by moving the right hip (ischial tuberosity) away from your heel, and push your heel into the floor. Repeat 5 to 10 times and then change legs.

Focus On

- Opposing points: pelvis (hips) and heel.
- Back leg straight.
- Active kite.
- Back of your neck (nape) relaxed.
- Fluid movements.
- Gradually increasing the range of motion.

Work On

Stretching the lower part of the Superficial Back Line (hamstring muscles, gastrocnemius) and mobility of the ankle.

Discussion

Compared to the previous exercise, which focused on dynamically stretching the entire SBL, this movement concentrates on the back of the thigh (hamstrings) and on the calf muscle.

Reduce the Level of difficulty

If you are unable to touch the floor without bending your legs, place a raised platform under your hands (step, yoga blocks). Reduce the height of the platform under your hands when it becomes easier to adopt the position.

Change Stimulus

The starting position is the same as for the basic exercise. Raise your front (right) heel slightly and move it inward, thereby externally rotating your right foot. Lower your right heel to the floor and move the hips away, causing your knee to extend. Lift your right heel again and move it outward, thereby internally rotating your right foot. Completely straighten your leg. Now, start the movement not from your heel but from the head of the femur, rotating outward and then inward. Has anything changed? Continue to move your heel for the desired number of repetitions and then switch legs.

Discussion

I love this exercise because you can really feel the strategy of sliding between the muscle bundles of the back of the thigh (between the heads of the hamstrings). The details provided in the box (The Hamstrings) will help you to understand the reason for, and importance of, moving your heels. Increasing the internal and external rotation of the leg by making the muscles slide from one to another improves elasticity, flexibility, and strength. Does this not sound like much to you?

I would like to add a side note about this wonderful phrase "to slide" (or "to glide"). In a myofascial setting this is a key strategy because everything is connected yet also separate. It is important to understand and know how to use "to slide" (or "to glide").

For a variety of reasons, including incorrect posture, lack of movement, a sedentary lifestyle, injury, and training that is too intense and performed incorrectly, tissue adhesions (both fascial and muscular) can often form. These can reduce or even fully impede the range of motion, creating tension and imbalances in the rest of the body. By applying sliding and swinging movements, we try to stimulate this matted tissue, but we also have to respect the fascia by stimulating it with suitable tension: Stimuli that are too weak are not going to induce changes, while stimuli that are too strong will stiffen the fascia. The following example should help you to visualize tissue adhesions. Try putting two cashmere sweaters in the dryer on high heat for 60 minutes. When you open the door you'll have a nice surprise waiting for you: The two sweaters will have become one and will be completely matted and inseparable. Simply put, this is what happens under our skin if we misuse our body. So what are you waiting for? Let's get on with the exercises!

Work On

Sliding between the fascial and muscular layers of the hamstrings and gastrocnemius (Superficial Back Line), as well as the bordering adductors and iliotibial band.

THE HAMSTRINGS
The hamstrings are a group of three muscles (the short head of the biceps femoris is not considered to be part of the hamstrings because it is a monoarticular muscle that does not originate from the ischial tuberosity):

- Long biceps femoris muscle
- Semitendinosus muscle
- Semimembranosus muscle.

All the muscles share a common origin (ischial tuberosity), innervation (ischial or sciatic nerve), and function. The three heads that make up the hamstrings all bend the leg at the knee and straighten the hip.

- The long head of the biceps femoris also externally rotates the leg and thigh.
- The semimembranosus and semitendinosus also internally rotate the leg (with the knee bent), while they adduct and internally rotate the thigh.

Let's take a detailed look at these muscles.

Biceps Femoris (Lateral Hamstring)
- **Origin**. Long head: Ischial tuberosity posteriorly and lower part of the sacrotuberous ligament. Short head: Linea aspera of the femur and supracondylar ridge.
- **Insertion**. Head of the fibula, laterally. Lateral condyle of the tibia and lateral part of the deep fascia of the leg.
- **Action**. Bends and externally rotates the leg. The long head extends (straightens) the thigh and is involved in its external rotation.

Semitendinosus (Medial Hamstring)
- **Origin**. Ischial tuberosity, with a common tendon at the long head of the biceps femoris.
- **Insertion**. Proximal part of the medial surface of the tibia.
- **Action**. Bends and internally rotates the leg. Extends (straightens) the thigh and is involved in its internal rotation.

Semimembranosus (Medial Hamstring)
- **Origin**. Ischial tuberosity, laterally and proximally to the origin of the biceps femoris and semitendinosus.
- **Insertion**. Medial condyle of the tibia, posteromedially.
- **Action**. Bends and internally rotates the leg. Extends (straightens) the thigh and is involved in its internal rotation.

Strategy

Dynamic stretching

Starting Position

Stand up straight, with your feet hip-width apart. Keeping your legs straight, roll your spine forward until your hands are touching the floor. Take three steps forward with your hands, place them shoulder-width apart and spread your fingers (each finger has its own function), and push your hands gently downward and forward (without moving them). In so doing, you should feel the Back Arm Lines activating (particularly the bottom of the trapezius, which connects to the coccyx). Push your shoulder blades toward your buttocks. Stretch your spine, keeping your heels firmly on the floor.

Movement

Walk in place, alternating your feet: As you raise your right heel with your forefoot on the ground, push your left heel into the floor. Continue walking in place for the desired number of repetitions.

Focus On

- Opposing points: hands and shoulder blades, heel and top of the head.
- Active kite and Arm Lines.
- Stretching your spine.
- Keeping your head between your arms.
- Pushing your heel into the floor.

Work On

- Stretching the gastrocnemius muscle and hamstring muscles (of the Superficial Back Line) and mobility of the ankle.
- Strengthening the muscles of the upper part of the SBL.

Discussion

In the Camel position, I often find that people raise their shoulders toward their head and bring their shoulder blades closer together, flattening the thoracic spine.

Let's see how the correct activations could help. They may be small details but that is not to say that they are any less important. In fact, they are essential.

Let's check. In the camel position, exert gentle pressure on the outside edge of your hands, including the pinkie. Can you feel the triceps and rhomboids activating? (Every finger has its own function.) If you can't feel it, try again with your elbows facing outward and an active kite. This will activate the Deep Back Arm Line. Now apply gentle pressure with your thumb, which is connected to the Deep Front Arm Line, and you may feel the pectoralis minor. Distribute the pressure across these two points to balance the two lines out.

In this and many other positions, we often lose such connections by using usual or more comfortable motor patterns. Let me give you an example: When you raise your shoulders and shoulder blades toward your head, you use your upper trapezius, you internally rotate with your shoulder, your pectoralis minor shortens, and so on, resulting in pain of the cervical spine and shoulders (it is always a choice or a habit that determines which myofascial lines you tend to use).

Reduce the Level of Difficulty

If you cannot touch the floor with your hands, place a platform under your hands. Reduce the height of the platform gradually as you train, to generate new stimuli.

Change Stimulus

- Adopt the Camel position. Lift both your heels off the ground at the same time, bring them together, and lower them back to the floor. Lift both your heels together, move them outward, and put them down on the floor again. Repeat these movements for the desired number of repetitions.

- Camel with lateral movements: Adopt the Camel position, bring your feet together, and move your whole body to the right and then to the left. As you move you should feel the LL stretching. Keep your hands on the floor at all times.

- Change the starting position of your feet. Separate your heels but keep your toes together (internal rotation of the leg). Move to the left and right, either with your toes apart and your heels together or with your legs hip-width apart and your toes pointing forward. Stimulate, stimulate, stimulate.
- Dynamic camel. Adopt the Camel position and bend your knees toward the floor. Push your heels toward the floor and move your hips toward the ceiling to return to the starting position.

Strategy

Dynamic stretching

Starting Position

Get on all fours with your knees hip-width apart and your hands shoulder-width apart. Move your left foot to the outside of your left hand. Turn your toes slightly toward your left hand (slight internal rotation of the foot). Ensure that all your fingers and toes are on the floor, as well as the three contact points of the foot (head of the first and fifth metatarsal and heel).

Movement

Move your left hip away from your heel by extending your left leg in a fluid movement. Keep your foot firmly on the floor (this is important). Then move your left knee forward, beyond your toes, before repeating the entire movement.

Repeat as often as desired before moving on to the circles. Move your left knee forward, like in the movement just performed. Make a circle with your pelvis around your left heel and extend your leg again. Continue until you have completed 5 circles and then change direction for another 5 circles. Switch legs.

Focus On

- Fixed point: foot; mobile point: hips.
- Keeping your front foot firmly on the floor.
- Performing fluid movements.
- Starting with small movements and increase with each repetition.

Work On

- Stretching and sliding the SBL and part of the SFL (tibialis anterior), DFL/SL.
- Functional mobility of the hip.

Change Stimulus

- Get on all fours and place your left foot between your hands with your toes facing forward.

- With the same starting position as before, extend your left leg and place your heel on the floor with your toes pointing toward the ceiling. In this position, simply push your front heel into the floor with slight traction toward your pelvis, thereby including isometric work. Externally and internally rotate your thigh (head of the femur), causing your toes to move outward and then inward. You should really feel sliding between the femoral heads.

Move forward and backward as described in the basic exercise. This will further stretch the SBL.

- Change the position of your foot: Perform all the movements described with your right foot in hammer position. This will further activate the SBL.

Strategy

Active dynamic stretching

Starting Position

Get on all fours with your knees hip-width apart and thighs perpendicular to the floor. Place your hands shoulder-width apart and your arms perpendicular to the floor, and spread your fingers.

Movement

Move your left foot to the outside of your left hand in one fluid movement then return to the starting position with an equally fluid movement.
Repeat with the right leg and continue alternating for the desired number of repetitions.

Focus On

- Active kite.
- Fluid movements.
- Starting with small movements and increasing with each repetition.

Work On

- Stretching the SFL/DFL.
- Functional mobility of the hip.

Increase the Level of Difficulty

- Perform the basic exercise with your knees raised and your pelvis moved toward your heels. Begin with slow movements before increasing the speed. As you move your pelvis backward, imagine that you are loading a slingshot to throw a stone forward. This movement involves dynamic stretching but also includes the ENERGY strategy (see page 227).

- Get into the plank position with your arms straight. Move your right foot to the outside of your right hand, return to the starting position and repeat on the left side. Continue alternating.

- Keep your right leg forward and imagine that you want to pick something up off the floor in front of you with your right hand. Repeat the movement with your left arm. Repeat a few times and then change legs. Now try and move the object on the floor slightly to the left when you reach for it with your right hand. This will increase the sideways leaning of your upper body, thereby involving the Lateral Line.

- Keep your right leg in front of you and try to lay your right forearm and then your left forearm on the floor. You should feel the stretch on your left quadriceps, right inner thigh, and back increase.

- Keep your right leg in front of you and raise your right hand, turning your upper body and following your hand with your eyes. Return to the starting position and repeat on the left side. It becomes a static stretching exercise for the legs. For the upper body it becomes a dynamic stretching exercise that works the Lateral and Spiral Lines.

- From the plank position, turn so that your left side almost touches the floor then return to the plank position quickly, and in an elastic and controlled manner without stopping in the twisted position, and repeat on the opposite side. This dynamic stretching movement will elastically work the Spiral Line.

6.8.2. Fascia Stretching of the SBL Against the Wall

Three Feet in Motion Against the Wall

Strategy

Dynamic stretching of the SBL against the wall

Starting Position

Get on all fours in front of a wall, with your knees hip-width apart and your hands shoulder-width apart. Move your left foot between your arms, and place your forefoot against the wall and your heel on the floor. The more you lift your forefoot, the more you increase the stretch on the sole of your foot, plantar fascia, flexor digitorum brevis, calf muscle, and hamstrings.

Movement

Move your left hip away from your heel by extending your left leg in a fluid movement. Keep your forefoot against the wall. Return to the starting position by moving your left knee toward the wall. Then repeat the entire movement.

Repeat as often as desired and then switch legs.

Focus On

*   Fixed point: foot; mobile point: hips.
*   Keeping your forefoot against the wall.
*   Straight back.
*   Active kite.

Work On

*   Stretching and sliding of the SBL (Superficial Back Line).
*   Functional mobility of the hip and ankle.

Change Stimulus

*   Slightly internally or externally rotating the foot that is against the wall is enough to change everything.
*   Transform the exercise into active isometric stretching of the agonist/antagonist. Agonist: Push your forefoot gently against the wall. Antagonist: Lift your forefoot slightly away from the wall.

Spreading Your Toes

Strategy

Fascia static stretching of the SFL

Starting Position and Movement

Stand up straight and lean against a wall to maintain your balance. Move your big toe forward and all your other toes backward. Hold the position for a few breaths.

Still in a standing position, move your big toe backward and your other toes forward (phalanges). Hold the position for a few breaths.

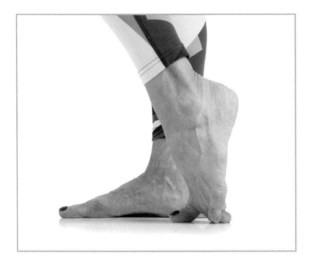 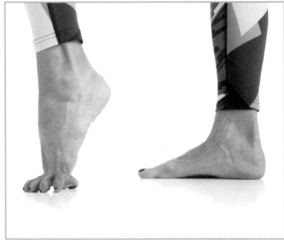

Focus On

- Not just bending the first phalanx of the toes but the entire mobile part to the proximal phalanges.
- Your heel must be perpendicular to the floor.

Work On

Stretching the SFL/SBL/DFL.

Change Stimulus

Reduce the level of difficulty by sitting down to perform the exercise, thereby diminishing the load, or use a platform to reduce the range of motion.

Discussion

I realize that it isn't an exercise for everyone, but if you try it and modify it correctly, you will notice both visual and sensory improvements (less pain) after just a few repetitions.

Strategy

Dynamic stretching of the SFL

Starting Position

Adopt the Camel position.

Focus On

- Mobile points: foot–knee; fixed point: hands.
- Active kite.
- Expansion.

Work On

Static stretching of the SFL/SBL.

Movement

Raise your right leg as high as you can into the air, keeping your pelvis in a neutral position.

From the hip, start to turn your thigh and pelvis. Move the little toe of your right foot away from the pinkie of your right hand. Bend your right knee, moving it as far away from your pelvis as possible. Return to the starting position and repeat with the other leg. Repeat 3 to 5 times per side.

Crab

Strategy

Dynamic stretching of the SFL

Starting Position

Sit down with your legs bent and spread shoulder-width apart. Place your hands on the floor behind your back, with your fingers pointing outward, and straighten your back as much as possible.

Movement

Activate the kite, keeping your elbows pointing outward, and raise your pelvis just half an inch (1 cm) off the ground. In this position you should feel the lower trapezius activating. Then lower your pelvis to the ground. You will only feel this activation if the shoulder blades are pushed down (active kite). Repeat the exercise a further 5 times. Now raise your pelvis, forming a plank with your body.

Hold for 3 seconds before returning to the starting position. Repeat as often as required.

Focus On

- Active kite.
- Elbows facing outward

Work On

- Stretching the SFL/Arm Lines.
- Strengthening the muscles of the SBL/Arm Lines.

Discussion

This is a really useful exercise on so many levels, including

- stretching,
- abdominal strengthening,
- activation of the Arm Lines,
- mobilization of the scapulohumeral joint and hip, and
- distribution of forces.

In the warm-up, the positioning of the hands depends on the client's needs. Let me give you an example: For someone with kyphosis, I would initially position their hands pointing outward. Choosing to position the hands pointing inward toward the buttocks would risk closing the shoulders forward, while positioning the hands so that they point backward could be too intense and would risk compromising the correct technique.

Change Stimulus

Change the position of your hands to vary the stimulation on the Arm Lines.

Strategy

Static stretching of the SFL

Starting Position

Adopt the half-kneeling position. Sit on your left heel. Place your right foot flat on the floor in front of you. Bend the toes of your left foot, with your heel perpendicular to the sole. Keep your upper body straight and your hands on your knee. Hold the position for a few breaths and lift your pelvis. Repeat another 3 times and then change feet.

Focus On

- Forefoot on the floor: Place the heads of the first and fifth metatarsals on the floor.
- Check if your heel moves outward (if it does, concentrate more on the first metatarsal and try to keep it on the floor). If your heel moves inward, push more with the fifth metatarsal and try to align it.

Work On

Static stretching of the Superficial Front Line (SFL).

Increase the Level of Difficulty

- Tilt your upper body backward, keeping your back straight. You should feel the stretch on the sole of your foot and on your quadriceps increase.

- In this variation sit on both your heels, with your toes bent.

Tilt your upper body backward and place your hands on the floor for support. Stay in this position to statically stretch the plantar fascia and quadriceps or alternatively move back and forth, increasing the range of motion with every repetition, turning it into a dynamic stretching exercise.

- Sit on both your heels, with your feet straight. If you have difficulty sitting on your heels, perform the exercise on page 95 (FEEL IT strategy): It will really help you.

Lean back as described in the previous exercise. If you feel ready, lean back completely until your shoulder blades touch the floor. Keep your knees hip-width apart and as close to the floor as possible. Keep your lumbar spine in a completely neutral position. Pulling your lumbar spine upward and creating a hyperlordosis (arching of the lumbar spine) will compress the lumbar vertebrae, and therefore the intervertebral discs, resulting in tension or pain in the lower back.

Hold the position for a few breaths and try to eliminate any tension as you breathe.

Work On

Static stretching of the Superficial Front Line (SFL).

Discussion

Seen from the outside, it may seem easy to adopt this position, but I can assure you that it isn't. How many times do you allow your feet to move freely? They are always trapped in shoes of varying levels of comfort, and when you get the chance to free them, it is your feet themselves that are unwilling. Conclusion: Work correctly, starting with simpler exercises before increasing the level of difficulty. Listen to your body: When the exercise is too intense, the pain will hinder your breathing. That is a clear sign that it is not doing you any good and that you will not improve. If you work correctly, your feet will thank you for it. Working on your feet helps to

- improve your gait,
- improve stability,
- minimize knee and back problems,
- eliminate cervical (neck) pain, and
- improve athletic performance.

Strategy

Dynamic stretching of the SFL

Starting Position

Lie on your back, with your legs bent and hip-width apart. Place your hands behind your ears, with your fingers pointing toward your shoulders and your elbows pointing toward the ceiling.

Movement

Create pre-tension by pushing your heels and palms of your hands into the floor. Start the movement by pushing your knees forward toward your toes and raising your pelvis. Relax your buttocks and let your chest between your arms fall toward the floor.

Focus On

- Palms of your hands in contact with the floor.
- Elbows pointing upward.

Work On

Fascia stretching of the SFL/AL.

Discussion

This is an exercise to prepare you for the full bridge, which involves raising your upper body by fully extending your arms. The position of the hands makes this exercise interesting. Any compensations you may have on the Arm Lines come to the fore with this exercise, such as difficulty keeping your whole hand on the floor or elbows that point outward instead of upward.

VARIATION: Advanced Bridge

Starting Position and Movement

Sit down with your legs bent and hip-width apart and hold your ankles with your hands.

Keep your ankles straight and start to lower your back toward the floor until you are completely lying on your back.

While still holding on tightly to your ankles, start to push your knees forward by raising your pelvis. Hold the position for two breaths and lower your back to the floor to return to the starting position. Repeat as often as required.

Focus On

* Keeping your knees hip-width apart.
* Fixed point: upper part of the shoulder blades; mobile point: knees.
* Try to stretch and then to lift.

Work On

* Static stretching of the Superficial Front Line and DFL.
* Strengthening the muscles of the SBL.
* Functional mobility of the spinal column.

Discussion

I repeat! The goal is not to get as high as you can at the expense of correct technique. Precise movements down to the smallest detail will guarantee that you improve and change your fascia.

Let's look at the final bridge position. Remember that the goal is to stretch the SFL. It could be performed as follows: Raise your upper body by arching your thoracic spine, lift your shoulder blades completely off the floor, turn your elbows inward (external rotation of the arms), bring your shoulder blades together (it will help you to raise up even further), and move your knees apart. Looking at the execution, you might think that you are really flexible. But is this really the case?

Turning your elbows inward and bringing your shoulder blades together activates the muscles of the upper back (rhomboid, middle trapezius) and shortens the latissimus dorsi (origin and insertion move closer together). All this to lift your upper body a little higher.

This both stretches and shortens, increasing your range of motion, but your flexibility does not change and certainly does not improve. I often use the image of a blanket that is too short: If I pull one end to cover my shoulders, I expose my feet at the other end.

Try it this way: Keep your elbows pointing outward and move your shoulder blades slightly away from each other (this will enable you to maintain thoracic kyphosis). Try to move your thoracic spine away from your sternum (imagine the two mobile points moving away from each other). Keep your shoulder blades open. Push your knees forward and then try to lift your pelvis. The result is expansion and not compensation.

Effects: Stretches the SFL, the Arm Lines, the LL, and the DFL and creates volume in the rib cage.

Strategy

Static stretching of the SFL

Starting Position and Movement

Get into the plank position with your arms straight.

Breathe in and move your sternum (fifth rib) toward the ceiling, activate your shoulder blades, and let your pelvis fall toward the floor so that the shape of your body resembles a boomerang.

Hold the position for a few breaths, relax by lying on the floor, and repeat for the desired number of repetitions.

Focus On

- Fixed point: feet; mobile point: sternum.
- Active kite.
- Elbows facing outward.

Work On

- Static stretching of the SFL and DFL.
- Strengthening the muscles of the SBL.
- Balance.

Change Stimulus

Adopt the same position as in the previous exercise but this time with the tops of your feet flat on the floor. Hold the position and turn your head to the right and then to the left. Repeat as often as desired.

Statue

Strategy
Static stretching of the SFL

Starting Position and Movement

Get on all fours with your knees hip-width apart and your hands shoulder-width apart. Lift your left leg and hold your ankle with your right hand. Move your sternum (fifth rib) as far away as possible from your raised knee to stretch the Superficial Front Line. Hold the position for a few breaths and then change sides. Repeat as often as required.

Focus On

- Mobile points: fifth rib and knee.
- Active kite.
- Elbow facing outward.

Work On

- Static stretching of the SFL and Front Arm Line.
- Strengthening the muscles of the SBL.
- Balance.

Change Stimulus

Perform the same exercise but this time hold your right ankle with your right hand. The challenge is not to lose your balance.

6.8.4. Fascia Stretching of the SFL Against the Wall

Bridge With Your Back to the Wall

Strategy

Static stretching of the SFL

Starting Position

Stand up straight and lean your sacrum, thoracic spine, and head against the wall, with your legs shoulder-width apart, your arms down by your sides, and the palms of your hands against the wall.

Focus On

• Fixed point: thoracic vertebra (T12); mobile point: hands.
• Expansion.
• Active kite.
• Elbows shoulder-width apart.
• Place the whole of the palm of your hands against the wall.

Work On

Static stretching of the Arm Lines, DFL/SBL/SFL.

Discussion

This exercise is excellent preparation for the full bridge.

Increase the Level of Difficulty

With your hands against the wall next to your ears, start to walk forward slowly, creating an arch from your feet to the top of your head. Feel your superficial front line stretch. Expand in all directions: hands toward the wall, feet and lumbar spine toward the floor, pelvis forward, sternum backward. Imagine that you are a balloon being blown up. If you feel compression in your lumbar spine, it is because you are compressing the lumbar vertebrae. In this case, go back slightly and stretch, creating space in this region. Better?

Movement

Raise your arms, bend your elbows, and place the palms of your hands on the wall next to your ears, with your fingers pointing downward and your elbows shoulder-width apart. Hold for a few breaths and then return to the starting position. Repeat as often as required.

Bird Dog Against the Wall

Strategy
Feeling everything

Starting Position and Movement
Get on all fours an arm's length from the wall, with your knees hip-width apart and your hands shoulder-width apart. Ensure your back is in a neutral position, raise your right arm, and put your right hand on the wall, at shoulder height, with your fingers spread. Hold the correct position for 30 seconds and then switch arms. Repeat 3 times per arm.

Focus On
- Active kite.
- Active shoulder blades: Push your right shoulder blade towards your left buttock (FL).
- Externally rotate your humeral head (arm); elbow pointing outward.
- With the back of your feet on the ground, push them gently into the floor (SFL).

Work On
- Feeling and stretching the Arm Lines.
- Stretching the SFL/DFL.
- Strengthening the SBL/FL/AL.

Discussion
This great exercise is useful in so many ways. Despite its apparent simplicity, it is very difficult to do, particularly when it comes to holding the correct position. All my clients (whether young or old, sporty or athletic) do this exercise, and I promise you that although there may not be tears, there is certainly sweat!
Here are a few reasons to improve:

- Breathing – expansion in the thoracic-lumbar region improves posture.
- SFL/DFL – move the fifth rib away from the pubis.
- Stretching and strengthening of the SFL/SBL – the two lines balance each other out.
- Stretching and strengthening - the correct position balances the front and back arm lines and strengthens the triceps and the latissimus dorsi.
- Feeling the correct position of the arms (external rotation of the humerus, internal rotation of the radius), the shoulder blades, and the thoracic kyphosis.
- Preparation for an overhead press.
- Work on a closed chain.

Let's reflect for a minute. Is this not a core exercise? I would say that it is. The entire rib cage is involved. We have to go beyond the fashion of merely training the abdominals with standard crunches. Remember: Expand rather than contract.

Change Stimulus
- Perform the exercise with your feet in hammer position.

Increase the Level of Difficulty

- Adopt the same position as the basic exercise but this time place both your hands on the wall. Hold the correct position for 30 seconds and repeat 3 times.

- Place both your hands on the wall and raise one leg. Hold the correct position for 30 seconds and then switch legs.

- Adopt a dog posture (on all fours with your knees raised), and place one hand on the wall. Hold the correct position for 30 seconds and then change sides. Repeat 2 or 3 times per side.

- Place both your hands on the wall. Without moving your pelvis or arms, slowly slide your right foot backward until your leg is completely straight. Keep your spine in a neutral position. Then slowly extend your left leg until your body is suspended, creating a continuous line from your heels to the top of your head. Push the bases of your first metatarsals gently into the floor and you will instantly feel more stable. Hold the position for 30 seconds. Repeat as often as required.

VARIATION 1

Change direction by turning 180 degrees so that your feet are next to the wall. Get on all fours, far enough away from the wall that you can straighten your right leg. Place your right foot flat against the wall. Push your right foot gently against the wall and simultaneously move your sternum (fifth rib) in the opposite direction. You should feel the SBL activating (hamstrings, spinal erectors) and the SFL stretching. Hold the position for a few breaths and then switch legs.

VARIATION 2

A more advanced version of this exercise involves raising your left arm straight out in front of your head, with your hand in a fist. The fist will give you greater stability. Everything else is as per the starting position.

VARIATION 3

Instead of raising your arm, try placing both your feet (first one and then the other) on the wall, hip-width apart and at shoulder height. Expand your chest toward the ceiling. Keep the thoracic kyphosis intact and move the sternum (fifth rib) forward, away from your toes. Elbows face outward. It is a position that requires extraordinary force distribution throughout the whole body, and you will find it extremely difficult if you try to use individual muscles or load all your weight on your arms or hands. Expand your upper body in all directions (like an inflatable mattress), and don't forget your hands: The palm of your hand is connected to the latissimus dorsi.

Strategy
Dynamic stretching

Starting Position
Lie on your back, with your head next to the wall and the palms of your hands flat on the wall, with your fingers pointing downward. Bend and lift your legs, keeping your heels together. Keep your spine and pelvis in a neutral position.

Movement
Push the palms of your hands gently against the wall and simultaneously push your shoulder blades toward your buttocks, breathe in, and keep T12 on the floor. Hold this position and try and feel the connection from the palms of your hands (as you push gently against the wall) to your pubis and from your shoulder blades to your coccyx.

Maintaining all these connections, push your heels together gently and straighten your legs in front of you, keeping them together. Return to the starting position in a quick but controlled motion. Note: Keep your spine in a neutral position throughout. Repeat as often as required.

Focus On
- Opposing points: hands and shoulder blades (they move away from each other), hands and feet (they move away from each other).
- Fixed point: thoracic spine (T12) on the floor.

Work On
- Elastic stretching of the hip flexors and SFL.
- Feel the connections between the hands and pubis, the hands and coccyx, and the iliopsoas and feet.
- Strengthening the core and the SFL/SBL/DFL/AL.

Reduce the Level of Difficulty
Only do the first part of the exercise.

Discussion
This is an exercise that works on many levels, such as feeling, stretching, and strength, depending on how you want to use it.

If you straighten your legs and quickly return to the starting position, you will benefit from the elastic recoil of the hip flexor. In fact, from the moment you extend your legs, energy is stored (eccentric work on the hip flexor), and at the same time your core works (and I mean really works!) and stabilizes, but always keep your back in a neutral position.

6.8.5. Fascia Stretching of the LL/SL

Bamboo

Strategy
Dynamic stretching

Starting Position

Stand up straight, with your legs together. Raise your left arm above your head and keep your shoulder blade pushed toward your buttocks.

Movement

With a fluid movement, slide your pelvis to the left, creating an arch with your whole body. Push the outside edge of your left foot into the floor, moving the pinkie of your left hand in the opposite direction and creating tension on that whole side of the body. Return to the starting position. Repeat as often as required and change sides.

Focus On
- Active kite.
- Fixed point: foot; mobile point: pinkie.

Work On
- Dynamic stretching of the entire Lateral Line.
- Hip mobility.

Discussion

The bamboo plant is an excellent way to think about our movement. Bamboo is pliable and resilient, elastic and elegant, a combination of refinement and plant beauty. It is how our body should aspire to be.

Change Stimulus

- To begin, perform the movement slowly and in a controlled manner. Once you have mastered the movement, start to vary the speed. Move to one side, creating tension on the entire Lateral Line, and then return elastically to the starting position without apparent effort (imagine pulling a bow to fire an arrow). As you tilt your upper body, put your right foot behind your left leg, increasing the stretch on the LL. Return to the starting position with your feet together and then, as you tilt your upper body again, move your right foot in front of your left leg. At the beginning, it is best to hold the stretch for a second, gradually making the movement more dynamic as your technique improves. Crossing your right foot in front of your left leg stretches the rear of the LL (gluteus maximus) at the height of the pelvis. Crossing your right foot behind your left leg stretches the front of the LL (tensor fasciae latae) at the height of the pelvis.
- Hold the position: It will become a static stretching exercise.

Increase the Level of Difficulty

As you lean, move your hand slightly diagonally and in front of you (your upper body will follow). This will increase the stretch on the rib cage. Repeat from the beginning, turning slightly backward.

Strategy

Active dynamic stretching of the LL

Starting Position

Stand up straight. Spread your legs hip-wldth apart, with your arms down by your sides.

Movement

Stretch your arms out in front of you with your hands together. Slide your pelvis to the left (coronal plane) and slightly bend your legs. Return to the starting position and repeat on the other side. Repeat as often as desired.

Discussion

This exercise really does give a full-blown *wow* factor. Perform the following test to see if this simple exercise could improve your flexibility and mobility.

Stand up straight, with your feet hip-width apart. Lean your upper body forward and try to touch the floor with your hands, keeping your legs straight. Make a mental note of the distance between the floor and your hands and the tension on the back of your thighs and back.

Now correctly perform the Shift Squat exercise 6 to 10 times per side. Then repeat the test by leaning forward once again.

Has anything changed? Has your ROM increased? Do you feel less tension? If so (and I'm sure the answer is yes), let me ask you a question: If you can improve your ROM with such a simple exercise, why not use it to warm up before doing squats or the Pilates Roll Up exercise?

You may think that I am some kind of magician, but don't mistake me for Merlin, who can perform miracles at will. All joking aside, you may be wondering what is behind this instant improvement. The sideways movements promote sliding between the various myofascial lines (LL/SFL/SBL); the Lateral Line affects the movement of the thigh (quadriceps and hamstrings, SFL/SBL). The iliotibial band (fascia) acts as a hydraulic amplifier around the thigh because its inner fibers run horizontally around it. And it must be mentioned that almost all the myofascial lines pass through the pelvis except the Arm Lines.

At this point I don't have anything further to add -except FReE...

Focus On

- Fixed points: feet; mobile point: hips.
- Knees aligned (your pelvis slides on the coronal plane).
-

Work On

- Activating, sliding, and stretching the Lateral Line.
- Hip mobility.

Change Stimulus

Perform the same sideways movement with your pelvis but this time raise your arms. This will increase the involvement of the upper Lateral Line, the Arm Lines, and the upper body.

The entire exercise can also be performed against the wall (see page 204).

Increase the Level of Difficulty

- Shift into half-kneeling position: Move sideways in half-kneeling position. Repeat 6 to 10 times per side and repeat from the beginning, changing the position of your legs.

- Shift split squat: To increase instability, perform the sideways movement with your back knee raised off the ground in a lunge position.

Strategy

Dynamic stretching

Starting Position

Stand up straight and cross your legs (right foot behind), keeping your feet close together. Bend your upper body forward until your fingers touch the floor.

Movement

Bend your left knee, keeping your foot flat on the floor, and move your pelvis to the right. Return to the starting position, straightening both legs. Complete the desired number of repetitions using fluid movements and then change sides, crossing your left leg behind your right, bending your right knee, and moving your pelvis to the left.

Focus On

- Fixed point: feet; mobile point: pelvis.
- Keep your feet flat on the floor (three contact points).

Work On

You should clearly feel the Lateral Line of your straight leg stretch.

Discussion

Sorry for showing you my behind, but it is the best way to explain the exercise to you. It may seem quite an intense stretching exercise at first, but you should soon see an improvement.

Change Stimulus

Use a platform (step, box) to reduce the level of difficulty if you are unable to touch the floor with your hands.

Slowly move your pelvis and return in an elastic but controlled motion to the starting position.

Windmill in Half-Kneeling Position

Strategy
Dynamic stretching of the LL

Starting Position

Adopt a half-kneeling position (left foot and right knee on the floor), with your arms outstretched on either side of your body and the palms of your hands facing downward.

Movement

Bend your upper body to the right, raise your left arm above your head, and lower your right hand to the floor.

Focus On

- Active kite.
- To return to the starting position, start the movement from your ribs on the left side and not from your arm. This will increase the sliding and stretching of your chest (internal and external oblique muscles, intercostal muscles).
- Keep your arms straight so as not to lose the connection with your Arm Lines.

Work On

- Flexibility and elasticity of the LL/DFL, iliopsoas.
- Hip mobility.

Change Stimulus

Perform the same movement but this time also rotate your chest. Start by raising your sternum (fifth rib) upward and then turn your chest. Bring your right hand toward your right heel, return to the starting position, and then turn your chest to the left toward your left heel. Try to stretch and not to compress: "Create an expansion". You should not feel tension in your lumbar spine. If you do, reduce the range of motion and concentrate on stretching.

Make everything more elastic (slow descent, elastic ascent) or use the domino effect to return to the starting position and move your upper arm forward and downward, thereby raising your upper body.

Strategy

Dynamic stretching of various lines: LL/DFL/SL/AL

Starting Position

Adopt a half-kneeling position with your left leg stretched out to your side. Align your left heel with your right knee, with your left foot facing forward.

Movement

Breathe in: Stretch and bend your upper body to the right, raise your left hand, and move it away from the outer part of your left foot, thereby stretching the entire left side of your body. To return, stretch up through the left hand as you move your left arm toward the ceiling, bringing your upper body upright. Breathe out and return to the starting position. Repeat the entire sequence 3 times.

When leaning and raising your arm, the quality of movement is important in order to strengthen the muscles of the upper part of the LL (concentric work on your right side and eccentric work on your left) and to stretch the lower part (from your pelvis to the outer edge of your foot).

In the return phase the recoil quality becomes important: the elastic recoil of the fascia.

On the fourth repetition, place your right hand on the floor as you tilt your upper body. Hold this position and stretch as much as you can. Start to turn your upper body toward the floor while moving your left hand as close as possible to your right hand on the floor. Raise your arm to return to the starting position. Repeat 3 times and then change sides.

When rotating, focus on the sliding of the fascia (latissimus dorsi, serratus anterior, obliques).

Three-Legged Windmill

Strategy
Dynamic stretching of the LL/SL/DFL

Starting Position
Get on all fours with your hands shoulder-width apart and your left leg stretched out to the side. Align your left heel with your right knee, with your left foot facing forward. You could align your right knee with your hip, but if you want to increase the stretch on the DFL, move your knee outward slightly.

Movement
Raise your left arm toward the ceiling by turning your upper body. Then turn your upper body toward the floor, placing your left hand on the right side of your ribs. Bend your left knee and right elbow at the same time, moving it outward. Lower your upper body as close to the floor as possible while looking at your right elbow. Return to the starting position.

Repeat with your right arm. Complete the whole sequence twice on each side and then change legs and repeat from the beginning.

Strategy

Static stretching

Starting Position

Get on all fours (using your forearms for support rather than your hands), with your knees hip-width apart and your hands shoulder-width apart. Extend your right leg behind you until it is completely straight.

Movement

Start to externally rotate your right thigh (femoral head, external rotation of the leg) until your pelvis starts to turn slightly, keeping your shoulders parallel to the floor. Hold the position for a few breaths and then change sides.

Focus On

- Fixed point: elbow.
- Active kite.

Work On

Static stretching of the LL/SL.

Boomerang in Sideways Half-Kneeling Position

Strategy

Dynamic stretching of the LL/SL/DFL/AL

Starting Position

Adopt a half-kneeling position, with your left foot out to your left side. Align your left heel with your right knee.

Movement

Move your left knee toward your toes and move your right arm in the same direction by turning your upper body. Keep your left foot firmly on the floor as you move your pelvis to the right. Bring the right arm overhead so that your right side arches to the left. Remember, it is the outward movement of your pelvis that stretches your left leg.

Repeat the entire sequence for the required number of repetitions, without stopping in the starting position, and then change sides.

Focus On

- Fixed point: foot; mobile point: pelvis.
- Pelvis in a neutral position.
- Active kite.

Work On

- Stretching of the Lateral Line, Deep Front Line, Spiral Line, and Arm Lines.
- Hip mobility.

Discussion

This is a dynamic, fluid, and elegant exercise. Imagine throwing a boomerang: Load by tilting to the side and then let go, bending your leg.

Change Stimulus

- For static stretching, lean to the side and hold for a few breaths or add foot traction for isometric stretching.
- Adopt the same position as described above but this time move your right foot inward. This will increase the stretch on the DFL.

Strategy

Dynamic stretching of the LL

Starting Position

Get into the plank position with your arms straight.

Movement

Bend your left leg and place your foot between your hands, keeping your right leg as straight as possible. Breathe in and move your left knee outward, placing the outside edge of your foot on the floor. Breathe out and bring your knee back to the middle. Complete the desired number of repetitions, slightly increasing the range of motion of your thigh with each movement. Repeat on the opposite side.

Focus On

- Fixed point: outside edge of the foot; mobile point: knee.
- Active kite.

Work On

- Dynamic stretching of the LL and part of the DFL.
- Hip mobility.

Discussion

This is a useful exercise for anyone who has a waddling gait (external rotation of the feet) or valgus knee.

Let's take a look in more detail: The outer part of the leg (lateral crural compartment, peroneal muscles, also known as the fibularis muscles) is stimulated. Some muscle groups must work harder, increasing the tension on the Lateral and Spiral Lines, which may give rise to other postural imbalances.

How can we improve? Start with a simpler movement, like the Crab in Squat Position on page 117. How do you know if you should increase the level of difficulty? When you can perform the exercise relatively easily, you are ready to increase the level of difficulty.

To align the limbs, it is not sufficient to only include stretching exercises of the DFL and hip mobilization. If you diligently follow a set routine, the change will be considerable in relatively little time. But it is not enough just to do the exercises because the movements should always be performed, for example, when you walk, when you are standing still, when you are climbing stairs, etc. You can check how your feet are positioned and which myofascial lines you are using at any time.

Change the tension and the bones return to their correct position. Use creates the shape, for better or worse.

Change Stimulus

- To simplify the exercise, start from a half-kneeling position or place your hands on a platform to reduce the range of motion (ROM).
- To increase the stimulation, hold the position with your knee turned outward for a few breaths. This will statically stretch the Lateral Line.

Increase the Level of Difficulty

Get into the plank position with your arms straight. Bend your left knee, place your left foot between your hands, and externally rotate your knee. Place the outside of your left leg completely on the floor and then extend your right foot. Try to maintain a 90-degree angle between your calf muscle and thigh. Hold the position. Breathe in and turn your head toward the toes of your left foot and then to the right. Complete the desired number of repetitions and then repeat with the other leg.

Change Stimulus

- Isometric static stretching – **agonist**: Hold the position, pushing the outside of your front foot gently downward onto the floor (without lifting your knee off the ground). This will activate the LL.
- Isometric static stretching – **antagonist**: Hold the position, slightly raising your front foot off the ground. This will activate the DFL.
- This is a good workout for the hip capsule.

Strategy

Isometric static stretching

To facilitate the Little Mermaid exercise on page 192, we are first going to mobilize the hip joint.

Starting Position

Sit in the "Z" position, with your left leg in front and your right leg turned to the side, stretching your spine.

Movement

Place your hands on your knees. Breathe in and push your left knee and foot toward the floor. Move your left knee slightly away from your left hand. Hold for 3 to 5 seconds, breathe out, and release the tension by gently pushing your hand toward the floor. Repeat 3 times.

Breathe in and push your right foot toward the foot by slightly raising your knee from the ground and pushing your right hand upward slightly. Hold the position for 3 to 5 seconds, breathe out, and release the tension, gently pulling your right knee toward the ceiling with your right hand. Repeat 3 times. Repeat from the beginning, changing the position of your legs.

Focus On

- Strength used: 5 to 20 percent.
- Active kite.
- Pelvis still.
- Breathe in as you exert pressure.
- Breathe out as you release the tension.
- Do not overdo it.

Work On

- Isometric static stretching of the hip rotators.
- Hip mobility.

Change Stimulus

- Position your legs at 90–90–90 degrees, thereby changing the angle of the hip. You will now be working on the end range of this joint by involving the joint capsule (see chapter 5).
- Add movement of the upper body: Raise your arms to the middle of your chest. With every repetition, as described before, breathe out as you move your upper body forward slightly toward your legs, thereby increasing the tension. Keep your back straight.

Little Mermaid

Strategy
Static stretching

Starting Position

Sit in the "Z" position, with your left leg in front and your right leg turned to the side, stretching your spine.

Movement

Raise your arms above your head and grasp your left wrist with your right hand, keeping your shoulders down. Breathe in, stretch, and lean your upper body to the right. Gently pull with your left hand to increase the stretch. Apply slight traction with your left hand and start to turn your sternum, rotating your upper body toward your right knee, thereby expanding your chest. Hold for two breaths and return to the intermediary position before returning to the starting position. Complete the required number of repetitions and then repeat with the other side.

Focus On

- Fixed point: left hip; mobile point: hands.
- Create expansion.
- Active kite.
- Breathing.

Work On

- Static stretching of the upper part of the LL/SL.
- Mobility of the thoracic spine.
- Strengthening the core muscles.

Change Stimulus

If the sitting position is uncomfortable, reduce the level of difficulty by sitting on a platform (medicine ball, step). Also see page 276: The ball could be useful.

Strategy

Dynamic stretching

Starting Position and Movement

From the same starting position described in the previous exercise, raise your right arm above your head and place your left hand on the floor next to your pelvis.

Stretch and lean your upper body to the left. Keep your left arm straight, creating an arch shape. Start the return movement from the ribs (your ribs should be the first things to move, rather than your arm), and all the rest (arm and head) then follows, with a domino effect. Repeat 3 times and then stop in sideways leaning position.

Turn your sternum, maintaining the expansion, and lower your right hand toward the floor. Keep your pelvis still. Return to a sideways leaning position before returning to the starting position. Repeat as often as desired.

Focus On

- Fixed point: left hip; mobile point: hands.
- Create expansion.
- Active kite.
- Breathing.

Work On

- Dynamic stretching of the upper part of the LL/SL.
- Mobility of the thoracic spine.
- Strengthening the core muscles.

Discussion

Let's look at this movement in detail: The first part of the exercise (raising your arm) is muscular, bending your upper body is myofascial stretching/sliding, and the return is an elastic and dynamic recoil effect (also see page 185).

Change Stimulus

- Stop in any position for a few breaths to make it a static stretching exercise.
- Perform fluid movements without stopping (when leaning) to make it a dynamic stretching and muscle strengthening exercise.

Increase the Level of Difficulty

- Stop in the position with your upper body turned, place your right hand on the floor, and stretch your back. Place your right hand under your left hand, keep your arms straight, and lower your chest toward the floor as you breathe in. Keep your back straight at all times (rotate around your central axis). Breathe out, stretch, and raise your upper body. Repeat as often as desired and change sides. This exercise both stretches and mobilizes your thoracic spine.

- Adopt the same position, with your hands on the floor and your arms straight. Turn to look at your left shoulder, looking for your right heel out of the corner of your eye. In this position, raise and lower your right leg (without it touching the floor). This exercise both stretches and strengthens.

Strategy

Dynamic stretching

Starting Position and Movement

Lie down on your back with your feet together, legs bent, and knees pointing outward. Place your hands next to your ears, with your fingers pointing toward your shoulders. Point your elbows toward the ceiling.

Create pre-tension by pushing the palms of your hands into the floor. Start the movement by raising your right knee and pushing it forward. Allow your pelvis to follow the movement by lifting and turning to the left. When in the bridge position, move your knee as far away as possible from your upper body.

Continue to push your hands against the floor. Lower your pelvis to the floor. Continue the movement by pushing your right knee outward (to the right), dragging your left knee with you as you move. Switch from right to left with fluid movements. Repeat as often as desired.

Focus On

- Fixed point: hands; mobile point: knee.
- Fluid movements.
- Active kite.

Work On

- Dynamic stretching of the SFL/DFL/SL/AL.
- Hip and back mobility.
- Strengthening of the muscles of the SBL.

Supine Corkscrew

Strategy

Dynamic stretching

Starting Position

Lie down on your back, with your legs straight and together, your arms bent above your head, and your fingers interlocked. Place your left foot (between the ankle and calf muscles – near the Achilles tendon) between the big toe and second toe of your right foot. Extend the toes of your left foot as much as possible.

Movement

Breathe in and turn your toes to the right, bringing your pelvis with you, while turning your head to the left at the same time. Breathe out and return to the starting position. Breathe in, turn your toes to the left (the range of motion on this side will be less), and turn your head to the right. Complete the desired number of repetitions and then repeat the exercise with your right foot between the big toe and second toe of your left foot.

Focus On

Fixed points: shoulders; mobile point: feet.

Work On

- Dynamic stretching of the LL/SL/DFL.
- Mobility of the hip, back, and toes.
- Strengthening of the muscles of the SBL.

Test

Once you have completed the repetitions with your left heel between the toes of your right foot, stand up and look at the space between your left big toe and second toe and between your right big toe and second toe. Listen to the support of both your feet and walk around the room. You should notice a difference between your two feet; there may be more space between your big toe and second toe, you should feel greater support from your right foot, and when walking you should feel greater support from your first metatarsal (slightly below the big toe). Repeat the entire exercise with your left foot.

Discussion

If you have completed the test, I'm sure you will agree that this exercise is very useful for improving foot support. This could have many benefits, including fewer knee problems, a more fluid gait, increased stability, and stronger legs. Conclusion: The better your foot support is, the better your performance will be (balancing foot support balances the forces). Think of the tripod of a video camera: If the three legs are not exactly the same height and equidistant from each other, the video camera will take distorted images or will fall over. The same is true of your body.

Strategy

Static stretching

Starting Position

Lie down on your back, with your legs straight and together, your arms outstretched at shoulder height, and the palms of your hands facing down.

Movement

Push your hands and heels gently down onto the floor, stretch and slowly lift your right leg, and move it to the left, crossing over your left leg, which must remain straight and on the floor. Continue the movement by lifting your pelvis and bringing your right leg toward your left hand. Hold for 3 breaths. Bring your leg back to the starting position and repeat with your left leg.

Focus On

- Fixed point: hands and shoulders; mobile point: foot.
- Keep the side of your raised leg long.

Work On

Static stretching of the LL/SL.

Supine Corkscrew in Action

Strategy
Active dynamic stretching

Starting Position
Lie on your back, with your legs straight and together, your arms outstretched above your head, and the palms of your hands facing each other.

From the prone position, pivot with your arms and back of your left foot, stretching and lifting your right leg and crossing it over your left leg. Your raised leg brings with it your pelvis, upper body, and then your arms until you are once again lying on your back.

Movement
As you push your left heel gently down onto the floor, stretch and slowly lift your right leg and move it to the left, crossing over your left leg that must remain straight and on the floor. Continue the movement, dragging your other leg with you. Your leg lifts your pelvis and then your upper body before reaching the prone position (lying on your front).

Focus On
- Your foot is the conductor that brings the rest of your body with it with a domino effect.
- Perform fluid movements.

Work On
- Dynamic stretching of the LL/SL.
- Muscle strengthening.

Discussion
Rolling exercises are my favorite – a game that reminds me of my childhood when I used to love rolling around and doing somersaults without limitations or pain and with such elasticity. What great memories! Perform these rolling exercises fluidly, elegantly, and loosely.

Increase the Level of Difficulty
Perform the entire exercise without using your hands: Raise your arms a few inches from the floor and perform the movement using only your legs. This requires control and coordination.

Strategy
Dynamic stretching

Starting Position
Lie down on your front and rest your forehead on your hands. Wedge your left ankle between the big toe and second toe of your right foot. Extend the toes of your left foot as much as possible.

Movement
Turn your heel to the right, bringing your pelvis with it. Go back to the starting position and turn your heel to the left. Complete the desired number of repetitions and then repeat the exercise with your right ankle wedged between the big toe and second toe of your left foot.

Focus On
Fixed point: shoulders; mobile point: foot.

Work On
- Dynamic stretching of the LL/SL/DFL.
- Mobility of the hip, back, and toes.
- Muscle strengthening.

Discussion
That's right! It's the same exercise as on page 196 but this time lying on your front.

Change Stimulus
Perform the exercise with your arms outstretched above your head.

Rotating Cobra

Strategy

Static stretching

Starting Position and Movement

Lie on your front and bend your arms in front of your head with your hands and feet together. Turn your upper body and head to the right, placing your left elbow on the ground in line with your right shoulder. Expand your upper body, raising your right elbow into the air and gently pushing your left elbow down into the floor. Hold the position for a few breaths and then change sides.

Focus On

- Active kite.
- Maintain the axial stretching of the spinal column.
- Feet together.

Work On

- Static stretching of the LL (upper part)/SL/DFL.
- Mobility of the spinal column.
- Stability and strengthening of the scapulo-humeral joint.

Discussion

You undoubtedly need a great deal of mobility to be able to perform this exercise correctly. The rotations on the thoracic spine are very interesting because they improve breathing and therefore increase drainage of the tissue.

Strategy

Static stretching

Starting Position

Lie down on your front (prone position), with your legs straight and together, your arms outstretched at shoulder height, and the palms of your hands facing down.

Movement

Stretch and slowly lift your right leg and move it to the left, crossing over your left leg, which must remain straight and on the floor. Continue the movement by lifting your pelvis and the right side of your chest off the floor, keeping your right shoulder close to the ground. Bring your leg back to the starting position and repeat with your left leg as often as desired.

Focus On

- Fixed point: shoulders; mobile point: foot.
- Keep the side of your raised leg long.

Work On

- Static stretching of the LL/SL.
- Mobility of the rib cage.

Change Stimulus

Increase the stretch on the Arm Lines (pectoralis major, biceps brachii, finger flexors, etc.). Once in the twisted position with your left leg raised, continue to turn your upper body, lifting your left arm into the air. With your weight on your right shoulder, try to move your sternum away from your shoulder, increasing the stretch on the pectoralis major. Return to the starting position and then change sides.

Screwdriver

Strategy
Dynamic stretching

Starting Position

Lie down on your left side, with your left leg straight and your right knee bent and in alignment with your pelvis. Keep your left arm straight and on the floor. Hold your ribs that are resting on the floor with your right hand. Use your hand to help turn your upper body. Stretch both arms toward the ceiling, at shoulder height.

Movement

Imagine holding a handle in each hand, one with an elastic band attached to the ceiling and the other with an elastic band attached to the floor. With your arms straight, breathe in, pull your right hand toward the floor, and push your left hand upward, thereby increasing the rotation of your chest. Breathe out and release the tension created in the rotation for elastic recoil. Repeat the exercise 3 times.

Having completed the required number of repetitions, breathe in and bring your right arm slowly past your head, thereby increasing the stretch on the upper body (LL). Keep your left arm raised, in line with your shoulder. Hold the position for a few breaths and return to the starting position. Repeat 3 times and then change sides.

Focus On

- Active kite.
- Small but controlled movements, gradually increasing the range of motion.
- Fixed point: pelvis; mobile points: arm and chest.

Work On

- Dynamic stretching of the Lateral Line and Spiral Line.
- Mobility of the thoracic spine and scapulo-humeral joint.
- Improved breathing and twisting.

Discussion

Upper body rotations against the floor or a wall are very useful because they limit compensations. The range of motion is limited, and in its place the efficacy of motion comes to the fore. At the same time, the contact points provide instant feedback. This exercise works on the rotations of both sides and is therefore useful for anyone who partakes in sports that involve rotations, such as golf, tennis, and volleyball.

Strategy

Dynamic stretching

Starting Position

Lie down on your left side, with your knees bent at pelvis height and your arms outstretched in front of you at shoulder height.

Movement

Breathe in, stretch, and then lift your right arm toward the ceiling. The movement of your arm brings your upper body with it, causing it to rotate (domino effect). Continue moving your arm as far as you can rotate your upper body without significant exertion. Return to the starting position, complete the desired number of repetitions, then switch sides.

Focus On

- Active kite.
- Appropriate movements within your ROM, increasing it gradually.
- Fixed point: pelvis; mobile points: arm and chest.

Work On

- Dynamic stretching of the Lateral Line and Spiral Line.
- Mobility of the thoracic spine and scapulo-humeral joint.
- Improved breathing and twisting.

Change Stimulus

Fan Style: Starting from the same position described in the basic exercise, place the fingertips of your right hand on the floor and start to move your hand toward and then slightly past your head. At this point, turn your hand so your palm is facing up. With your nails touching the floor, start to rotate your upper body. Finally, return to the starting position by performing the exercise in reverse.

6.8.6. Fascia Stretching of the LL/SL Against the Wall

Shift Squat Against the Wall

Strategy

Dynamic stretching of the LL

Starting Position

Stand sideways to the wall, about 1 1/2 feet away from it, with your feet hip-width apart and your arms straight and down by your side.

Focus On

* Fixed point: feet; mobile point: hip.
* Active kite.

Work On

* Activation, sliding, and stretching of the Lateral Line.
* Hip mobility.

Increase the Level of Difficulty

If you find it easy to touch the wall with your hip, move a bit further away from the wall.

Movement

Move your pelvis toward the wall and slightly bend your legs. Return to the starting position. Repeat as often as desired.

Change Stimulus

Rotating shift squat: As your hip touches the wall, turn your upper body to the left, involving the rotation of the upper body and the Spiral Line.

Also see page 181.

Strategy

Static stretching of the LL

Starting Position

Stand sideways to the wall, about 1 1/2 feet away from it, with your arms straight and down by your side.

Movement

Put your feet together. Raise your left arm above your head, placing your hand on the wall with your fingers pointing up. Place the palm of your right hand on the wall, keeping your arm down by your side. Move your pelvis to the left, away from the wall. Make an arch shape with your body, hold for 2 to 3 breaths, move your pelvis back to the middle, and repeat 3 times. Then change sides.

Focus On

*   Fixed points: feet and hands; mobile point: pelvis.
*   Active kite.
*   Keep your shoulders and hips aligned.

Work On

*   Static stretching of the entire Lateral Line and Arm Lines.
*   Functional mobility of the hip.

Change Stimulus

Bend your outer knee as you move your pelvis away from the wall.

Also see page 180.

Rotating Split Squat Against the Wall

Strategy
Active dynamic stretching

Starting Position
Stand up straight with your legs together. Stand sideways against the wall, with your pelvis brushing against it. Turn your upper body to the right, raise your outstretched arms to shoulder height and place them against the wall, and take a step backward with your left leg.

Movement
First flex your hip (moving the hip backward slightly). Slide down the wall slowly until your left knee touches the floor. Return to the starting position by pushing off with the heel and first metatarsal of your front foot. Repeat as often as desired and then change sides.

Focus On
- Active kite.
- Going down: First move the hip backward.

Work On
- Stretching the LL/SL/AL.
- Strengthening the muscles of the lower limbs and core.
- Functional mobility of the thoracic spine and shoulder.

Discussion

This can be thought of as a knee-saving movement.

Why do we flex the hip (moving the ischium backward) before we bend our legs? Because as you descend, this activates the hamstrings (muscles at the back of the thigh), which originate in the ischial tuberosity and insert into the tibia and the condyle of the fibula (see details on page 158). Moving the ischial tuberosity backward pulls the hamstrings as slight traction is applied to the tibia and fibula (SL), which both protects and stabilizes the knee. What happens if this activation is not performed prior to bending our legs? The weight moves to the quadriceps, and the knee moves forward, beyond the toes, working the SFL.

Results:

1. Many clients complain of knee pain.
2. The quadriceps (front of the thigh) develop much more.

When you come back up, you need to push your heels and the head of the first metatarsal against the floor, as well as move the ischial tuberosity backward to activate the hamstrings. Remember these activations the next time you go upstairs to protect your knees and tone your hamstrings; this is your homework.

You're working on a closed chain in this exercise. The interesting aspect compared to working with your own body weight is that your movement is impeded by the wall. At the same time, the wall also gives you instant feedback that helps you to understand your limits (putting your hands on the wall is a contact point that helps you to activate the myofascial lines of the arms and the core, making everything more effective and stable).

Reduce the Level of Difficulty

If the rotation of your thoracic spine is limited or you have compensations on the Arm Lines, move a bit further away from the wall to make the exercise easier.

Change Stimulus

Reverse split squat: Take a step backward with your right leg.

Test

Once you have completed this exercise on your right side, do this test.

1) Go for a walk: Do you notice a difference between your right and left sides?
2) Stand up, raise your right arm by leaning your upper body to the left, and then repeat on the other side.

Don't tell me you don't feel a huge difference in mobility, flexibility, and agility between the two sides.

Windmill Against the Wall

Strategy
Active dynamic stretching

Starting Position

Stand up straight with your toes against the wall. Stretch your right arm above your head, keep your left arm down by your side, place both your hands on the wall, and turn both your feet diagonally to the left.

Movement

Start the movement by pushing your left hip backward and to the side. At the same time, start to lean your upper body to the left. Continue to descend as far as you can with your legs straight and hold the position for a few seconds. Then push your right foot against the floor and slowly come back up until you are standing up straight. Repeat as often as desired. Finally, switch sides and perform the exercise again.

Focus On

- Fixed point: feet; mobile point: hip; opposing points: hands.
- Increase the range of motion slightly with each repetition.
- Keep your hands against the wall and your legs as straight as possible.
- Active kite.
- Straight (but not hyperextended) knees.

Work On

- Active dynamic stretching of the LL/SL/SBL/AL.
- Hip, thoracic spine, and shoulder mobility.

Reduce the Level of Difficulty

If you find it very difficult to lower your upper body, move away from the wall slightly. If the tension in the back of your thigh completely prevents you from descending, bend both your legs. Remember that the goal is to go down with your legs straight to stretch the lines involved in the legs and not to touch the floor at all costs.

Change Stimulus

Bend down slowly to increase the stretch, creating energy. Come back up in an energetic but controlled manner, using the elastic energy of the fascia.

Discussion

The Windmill is one of my favorite exercises. I recommend all possible forms and variations of this exercise because for me it expresses three-dimensional freedom.

There are few exercises that are as effective, complete, and elegant as the Windmill. This exercise is the perfect preparation for the bodyweight Windmill or the Windmill with kettlebells. When you try it, you will soon realize that the wall makes it more difficult because it prevents you from moving your upper body forward, but the trajectory of movement is perpendicular to the floor (which therefore increases the twisting of the upper body). In other words, it allows for very few compensations.

Having included the Windmill in my personal training and group sessions, I often see how difficult people find it to keep one arm straight up and turn their upper body without moving their arm forward (limitations on twisting the thoracic spine and arm extensions). Using the wall forces me to go against my natural tendencies while also providing two contact points, which help me to feel and better activate the myofascial lines of the arms.

Strategy

Dynamic stretching

Starting Position

Standing upright. Place your hands on the wall at chest height and slightly more than shoulder-width apart. Take two steps back and make a square with your body, with your upper body horizontal to the floor and your legs slightly more than shoulder-width apart. Put your head between your arms and keep your shoulder blades far apart from each other, maintaining the thoracic kyphosis.

Movement

Take your right hand off the wall and move it to the outside edge of your left foot as you start to rotate your upper body (sternum). Keep your left arm straight and your pelvis still as you turn your upper body. Hold the position for 2 to 4 seconds before turning your upper body back to the starting position. Repeat on the opposite side and complete the desired number of repetitions.

Focus On

- Fixed points: feet and pelvis; mobile points: sternum and hand.
- Active kite.
- Straight arms, palms of your hands against the wall.
- Straight legs.

Work On

- Dynamic stretching of the LL/SL/SBL/AL.
- Mobility of the thoracic spine.

Reduce the Level of Difficulty

Perform the Square Against the Wall With Plyo Box exercise (page 212).

Discussion

What do you see if you compare the Standing Square Against the Wall with the Standing Square Against the Wall With Plyo Box (page 212) exercises? You've hit the nail on the head! It's the position of the hands that makes the difference. Against the wall we increase the stretch on the Arm Lines, particularly the wrist (SFAL).

The connection only acts as a tensile mechanical connection when your arm is straight above your head.

Three-Legged Windmill Against the Wall

Strategy
Dynamic stretching

Starting Position

Get on all fours with your hands shoulder-width apart and your right leg stretched out to the side. Align your right heel with your left knee and place the outer edge of your right foot against the wall.

Focus On

- Fixed point: foot against the wall; mobile point: upper body and hand.
- Active kite.
- Stay on all fours.

Work On

- Dynamic stretching of the LL/SL/DFL/SBL.
- Stretching the sacrotuberous ligament.
- Mobility of the spinal column and hip.
- Stabilization of the shoulder blade.
- Improving breathing.

Discussion

As you turn your upper body toward the floor, imagine a chicken on a spit or rotisserie. This is exactly how you should turn, around the central axis of your spinal column. It is a rotation rather than flexion of the upper body. The initial tendency will be to move your pelvis toward your heel. Mmh... too easy! Your hips must always be aligned with your knees.

Change Stimulus

Perform the exercise on page 186.

Movement

Raise your right arm toward the ceiling by turning your upper body.

Then turn your upper body toward the floor, placing your right hand on the left side of your ribs. Bend your right knee at the same time as your left elbow, moving it outward. Lower your upper body as close to the floor as possible while looking at your left elbow. Return to the starting position.

Repeat with your left arm. Complete the whole sequence twice per arm and then change legs and repeat from the beginning.

6.8.7. Fascia Stretching of the LL/SL With Equipment

Elephant on the Plyo Box

Strategy

Dynamic stretching

Starting Position

Stand up straight about 20 inches (50 centimeters) from the Plyo Box, with your feet hip-width apart. Place your right heel on the Plyo Box with your knee slightly bent and your pelvis in the starting position. Keep your left leg straight, with your toes pointing forward toward the Plyo Box.

Movement

Straighten your right knee as you lean your upper body forward, moving your hands to the outside of your right foot. Continue the movement by moving your upper body, and therefore your hands, to the inside of your right foot. Continue these movements, increasing the range of motion a little with each repetition. Perform the desired number of repetitions and then switch legs.

Focus On

- Expansion.
- Keeping both legs straight.
- Moving your upper body, which will cause your arms to move.
- Bending your lumbar spine.
- Letting your head drop toward the floor.
- Maintaining the axis of the pelvis.

Work On

Primarily the flexibility and elasticity of the entire myofascial Superficial Back Line (SBL).

Discussion

Unlike the movement performed in the Elephant exercise (page 155), where you feel the stretching of your back and the back of your leg, in this variant you will feel the stretch concentrated on the hamstrings and calf muscle (back of the leg). This further highlights the importance of a simple concept: Keep changing to stimulate your fascia in as many different ways as possible.

Standing Square With Plyo Box

Strategy
Dynamic stretching

Starting Position

Stand up straight and place your hands, slightly more than shoulder-width apart, on a Plyo Box or on the backs of two chairs. Take two steps backward and lower your chest so that it is horizontal to the floor so that your body makes the shape of a square (with the wall and floor). Ensure your feet are slightly more than shoulder-width apart. Put your head between your arms and keep your shoulder blades far apart from each other, maintaining the thoracic kyphosis.

Movement

Take your left hand off the Plyo Box and move it to the outside edge of your right foot as you start to rotate your upper body (sternum). Keep your right arm straight and your pelvis still as you turn your upper body. In this position, move your left hand to the inside of your right foot and then back to the outside, gradually increasing the range of motion.

Perform the desired number of repetitions, return to the starting position, and repeat on the opposite side.

Focus On

- Fixed points: feet and pelvis; mobile point: hand.
- Active kite.
- Arms and legs straight.

Work On

- Dynamic stretching of the LL/SL/SBL/AL.
- Mobility of the thoracic spine.

Also see the exercise against the wall on page 209.

Strategy
Static stretching

Starting Position

Adopt a half-kneeling position half a step from the Plyo Box, with your left foot on the box, in line with your hip. Place your left hand on your left leg and stretch your right arm above your head. Expand your upper body upward.

Movement

With active kite (shoulder blades pushed toward your jean pockets), move your right thigh (or more specifically, the femoral head) slightly toward the Plyo Box, thereby increasing the stretch on the iliopsoas and rectus femoris. If you feel that you are unable to stretch beyond this position, hold the position and breathe. Otherwise, increase the stretch by expanding your upper body toward the sky, stretching and leaning slightly diagonally and backward. Imagine moving your rib cage away from your pelvis. Return to the starting position, repeat 3 times and then change sides.

Focus On

- Fixed point: knee; mobile point: fifth rib.
- Active kite.
- Expansion of the upper body.

Work On

Static stretching of the Superficial Front Line and Deep Front Line.

Discussion

During the exercise you should never feel compression of the lumbar spine, because this will cause you to compensate rather than to stretch.

Reduce the Level of Difficulty

Perform the exercise on the floor without a platform or reduce the height.

6.8.8. Fascia Stretching of the DFL

Gorilla

Strategy
Dynamic stretching

Starting Position
Stand up straight, with your legs slightly more than shoulder-width apart. Lean forward with your upper body, keeping your legs straight, and put your fingers under the soles of your feet.

Movement
Slowly bend your knees to lower your pelvis into a full squat, keeping your arms straight throughout. Extend your spine toward the ceiling, pushing your knees slightly outward. Hold the position for one breath. Push your head toward the floor as you straighten your legs to return to the starting position. Repeat as often as desired.

Focus On
- Keeping your arms straight at all times.
- Active kite.
- Spine in a neutral position.
- Knees in line with your toes (activation of the hip external rotation muscles).

Work On
- Dynamic stretching of the DFL and SBL.
- Hip mobility.

Reduce the Level of Difficulty
Place your hands on a platform in front of you (step, Plyo Box).

Strategy

Static stretching

Starting Position

Adopt the full squat position, place your elbows inside your thighs, and bring your hands together in front of your chest. Gently push your elbows against your inner thighs to increase the external rotation of your legs, and simultaneously straighten your spine.

Discussion

This is clearly not a suitable starting position for anyone with limited hip and ankle mobility. You could begin with simpler exercises, like the one on page 141.

Change Stimulus

- Dynamic frog: Perform the whole exercise dynamically, without stopping in the starting position or final position.
- Dynamic jumping frog: To make the exercise even more dynamic, put your arms straight out in front of you; perform a small jump with your legs, with your inner thighs pushing against your elbows; and jump back to return to the starting position.

Movement

Place your hands on the floor shoulder-width apart, bring your body weight forward, and keep your thighs and elbows together. Slowly transfer your weight onto your hands by lifting your legs off the floor. Hold for a few breaths and then return to the starting position to complete the desired number of repetitions.

Focus On

- Active kite.
- Elbows and knees pushed outward.

Work On

- Static stretching of the Deep Front Line.
- Strengthening of the stabilizer muscles of the arms and core.
- Balance.

Reduce the Level of Difficulty

Perform only half the movement. Move forward without lifting your legs off the floor. Hold this position and then return to the starting position, increasing the range of motion with each repetition. Once you feel comfortable performing this movement, try to lift one foot then lower it and lift the other foot.

Twisting Scorpion

Strategy
Dynamic stretching

Starting Position

Sit on both your heels, with your feet outstretched, your knees spread as far apart as possible, and your feet together at the toes. Put your hands on the floor in front of you, with your arms and spine straight. Try to pull your hands gently toward your pelvis to feel the activation of the kite.

Focus On

- Active kite.
- The foot in hammer position protects the knee.

Work On

- Stretching the psoas; the femoral head subsequently moves (DFL).
- Timing between movement of the ilium and the femoral head.

Discussion

This is a great exercise for stretching the psoas and subsequently moving the femoral head.

Movement

Lift your buttocks from your heels and move your pelvis back and forth. Repeat 4 times.

You are now ready to continue. Turn your pelvis to your left while simultaneously lifting your right foot off the floor (in hammer position). Move your right hand in front of your left hand. Turn your right foot outward as far as possible and internally rotate your femur. Activate the kite by pulling your hands slightly toward your pelvis. Stretch your sternum (fifth rib) upward and let your pelvis fall toward the floor. Hold the position for 3 seconds, return to the starting position, and repeat a further 3 to 6 times on the right side. Switch legs.

Change Stimulus

- As you turn your leg, slide the outside of your right hand forward to also stretch the latissimus dorsi.
- Isometric stretching: Hold the position for 5 seconds, pushing your right foot outward to increase the pull. Then relax in this position, increasing the ROM (internal rotation of the femur), and repeat another 3 times. It is useful to perform this exercise with a partner to provide light resistance.

Strategy

Static stretching

Starting Position

Lie on your back, with your legs straight and together, your arms straight and at shoulder height, and the palms of your hands face down.

Movement

Push your hands and your heels gently down onto the floor. Then lift your right leg toward the ceiling and move it slowly to the right, keeping your pelvis in a neutral position. Hold for 2 to 3 breaths. Bring your leg back to the starting position and repeat with your left leg.

Focus On

- Fixed point: pelvis; mobile point: foot.
- Keep the side of your raised leg long.
- Shoulder blades open.

Work On

Static stretching of the Deep Front Line.

Rolling Beetle

Strategy
Dynamic stretching

Starting Position
Lie on your right side, with your arms stretched straight ahead and legs bent.

Movement
Raise your left arm and leg and move them away from your right arm and leg. You should feel your abdominals activating and, at the same time, your inner thighs stretching. Keep turning until your left arm and leg are on the floor, allowing your right arm and leg to passively follow the movement. Perform the same movement with your right arm and leg to return to the starting position.

Focus On
- Turn the sacrum (keep your pelvis in a neutral position).
- Fluid and controlled movements.

Work On
- Dynamic stretching of the Deep Front Line and FL.
- Mobility of the hip and sacroiliac joint.
- Strengthening of the core and control.

Discussion
Move your upper arm and leg as far away as possible and lift your lower arm and leg as late as possible. This really is a domino effect. If you perform this exercise regularly, you will see real improvements in your deep squat position. It is also a fun exercise.

Change Stimulus
- Beetle stretch: Lie on your back and hold your big toes with your index and middle fingers. Ensure your arms are inside your thighs. As you breathe in, extend and spread your legs, keeping your back in a neutral position. You shouldn't feel any pain: Stretch only as far as you can keep your spine in a neutral position, then hold for 2 breaths, bend your legs again, and repeat from the beginning, carefully increasing the stretch.
- Perform the entire exercise with your head on the floor.

Increase the Level of Difficulty

- Transform the movements into gestures thanks to the rolling beetle that tries to get up: With your arms inside your legs and holding onto your big toes with your index and middle fingers, roll from one side to the other with controlled and fluid movements. Push your feet away from your chest while your hands pull toward your chest, creating an opposing force and increasing control of the roll.

- Do you want to make it even more difficult? Roll from one side to the other, like in the variation just described. When you roll to one side, straighten your upper leg and bend the other leg closer to the floor.

- To further increase the difficulty of the previous variation, when you roll to one side, keep rolling fluidly until you are sitting. Control your momentum to end up in a sitting position, pushing gently with your hand and foot. Keep practicing this exercise and your movements will soon become more fluid and controlled.

- Repeat the entire sequence 3 to 6 times per side and then stop in a sitting position. Cross your legs, lean your upper body forward and stretch your arms out in front of you, hold for a few breaths, and bring your breathing toward your lower back.

Discussion

Work on the same principles as in the Rolling Beetle exercise (page 218), but with this higher level of difficulty, a significant distribution of force is required, expressed as balanced and fluid movements. And this triggers a whole series of combinations to stand up, get onto all fours, and so on, expanding our motor patterns from a single movement up to the most complex of gestures. The sky is the limit.

6.8.9. Fascia Stretching of the DFL/AL Against the Wall

Splits Against the Wall

Strategy
Static stretching

Starting Position
Lie down on your back, with your legs straight and your buttocks against the wall. Keep your pelvis in a neutral position. Raise your outstretched arms to shoulder height, with the palms of your hands facing down and your elbows pointing outward.

Movement
Slowly spread your legs as far as you can while keeping your pelvis in a neutral position. Expand your rib cage and breathe toward your ribs that are on the floor. Hold for a few breaths. When you feel the tension diminish, relax and open your legs a bit more.

Discussion
In this exercise, the key to success is the position of your pelvis. In a position of pelvic retroversion (flattening the lumbar spine), the origin and insertion of the adductors come closer together, thereby inhibiting the stretch.

Focus On
- Neutral positioning of your pelvis, lumbar spine off the floor, thoracic spine on the floor.
- Active kite.

Work On
Static stretching of the Deep Front Line, Arm Lines, and SBL.

Reduce the Level of Difficulty
If you are unable to keep your pelvis in a neutral position, move your buttocks away from the wall until your pelvis is in the correct position. If you cannot keep your legs straight, bend your knees to make a diamond shape.

Strategy

Static stretching of the Arm Lines

Starting Position

Stand up straight with your back against the wall and with your pelvis and spine in a neutral position. Put your arms straight down by your sides, the palms of your hands against the wall, and your elbows facing outward.

Movement

Take one step forward, keeping your hands flat on the wall. Slide your hands up the wall. Internally rotate your humerus slightly and try to take another step forward, further raising your hands slightly. Hold the position for a few breaths then return to the starting position and repeat 3 times.

Focus On

- Opposing points: hands and sternum; create space between these two points.
- Avoid compensations.
- Slight internal rotation of the humerus, with your collarbones open and your elbows pointing outward.
- Active kite and your shoulder blades far apart from each other, maintaining the thoracic kyphosis.
- Imagine you have the spiderman/woman emblem on your chest and you are proud to show it to everyone but without compensations (like shoulder blades that are too close together, collarbones that close forward, palms or wrists that lift off the wall, etc.).

Work On

- Static stretching of the Front Arm Line.
- Strengthening the muscles of the Back Arm Line.

Discussion

This exercise literally drives me crazy because I put 80 percent of people who feel very strong and quite elastic with their backs to the wall. And that is where my enormous respect for this exercise comes from; it seems simple when you see it, but actually doing it is quite another story. But the great thing is that you feel your own limitations.

VARIATION 1: Sideways Spider Against the Wall

Starting Position

Stand sideways to the wall, about an arm's length away. Put your right palm on the wall, with your fingers pointing back. With your right arm slightly bent, twist your elbow back a little. In this position, activate your shoulder blade and latissimus dorsi and slowly straighten your arm.

Movement

Start the movement from your sternum, moving it away from your right hand. It doesn't take much to feel the tension in your arm increase. Hold for 3 seconds, relax, then repeat. Those who have not yet had enough can turn their whole body. Hold the correct position for a few breaths. Repeat twice more and then change arms.

Focus On

- Fixed point: hand; mobile point: sternum.
- Create space.
- Active kite.

VARIATION 2: Dancing Spider Against the Wall

Starting Position

Stand up straight, about 20 inches (50 centimeters) from the wall. Raise your right arm above your head and place your hand on the wall. Look at your right hand.

Movement

Let the dance begin!

Imagine having a dance partner who is holding your hand. Bring your head under your right arm by moving your left foot between the wall and your right foot and turning your whole body. (Opposing points: right hand and right foot. Create space between these two points.) In this position, continue to turn to the right with your pelvis. Hold the position and breathe. You should feel the tension in your chest and the outside of your right leg increase. Slowly move your pelvis back. Think of the movement of a mantis, which stays in one place by making small back-and-forth movements. Play with your lines, exploring where movements are more fluid and where they are blocked. In this position there are often blocks around the iliac crest, which I like to call "works in progress". Slowly clear the blocks by getting into the position and breathing. Increase the movement gradually as the tension abates.

Discussion

This is an insightful, effective, and fun exercise.

Only attempt this next step if you have successfully completed the last exercise and you are able to stabilize your shoulder blade and keep your humeral head in the correct position (screwing in a light bulb). Continue to turn to your left until your back is to the wall. Do not move your hand and keep your elbow straight. Hold the position for two breaths. You should feel your arm and SFL stretching without there being too much tension. If you feel too much tension, this is not the right exercise for you at this time. Reverse the movement to return to the starting position.

Return to the starting position. Move your right foot between the wall and your left foot by turning your body to the left. Keep looking at your right hand. (Opposing points: right hand and right foot.) Repeat from the beginning with the other side.

To sum up, let your partner drag you in all directions. There are only three rules: Keep your shoulder blades active, create space between the opposing points, and, of course, don't overdo it, because you don't want to end up with a dislocated shoulder.

Focus On

- Fixed point: hand; mobile points: feet, pelvis.
- Active kite.

Reduce the Level of Difficulty

Move closer to the wall and bend your elbow slightly.

6.8.10. Fascia Stretching of the DFL With Equipment

Sideways Modified Warrior on Plyo Box

Strategy
Dynamic stretching

Starting Position
Standing sideways and one step away from the Plyo Box, externally rotate your left foot and put your foot on the Plyo Box. Ensure that your left heel is in line with your right heel. Place your left hand on your left leg.

Movement
Push your left knee toward and beyond the toes of your left foot. Lift your right arm above your head and tilt your upper body slightly. Bring your left hand down to the Plyo Box and push your left elbow gently against your left inner thigh to keep your leg externally rotated. Return to the starting position and complete the desired number of repetitions, then switch legs.

Focus On
- Fixed point: foot on the floor; mobile point: knee of the bent leg.
- Pelvis aligned and no rotation of the upper body.
- Active kite.

Work On
- Isometric static stretching of the Deep Front Line.
- Static stretching of the LL and AL.
- Hip and ankle mobility.

Discussion
Before doing exercises that strengthen the leg muscles (squats, lunges, etc.), work on dynamic stretching, mobility, and activation of the abductors, adductors, flexors, and extensors of the pelvis. This will balance the tension felt in the pelvis.

Change Stimulus
Let's transform this dynamic exercise into isometric static stretching: Hold the position and push your knee gently against your elbow. Hold for 3 seconds then breathe out and increase the range of motion. Repeat 3 times and then switch legs.

Strategy

Dynamic stretching with weights

Starting Position

Stand up straight and take one step forward with your left leg. Pick up a weight in your right hand, put your arm up in the air, and push your shoulder blade down. Keep your spine in a neutral position.

Movement

First activate the kite then start to swing your left arm gently back and forth with small, fluid, and controlled movements. Slowly increase the range of motion. Repeat as often as desired and change sides.

Focus On

- Active kite.
- Straight arm, elbow pointing outward.
- Small and controlled movements.
- Keep your spinal column in a neutral position.

Work On

- Dynamic stretching of the Front Arm Line (pectoralis minor) and SFL.
- Strengthening the Back Arm Line.
- Improving breathing.
- Improving posture.
- Improving movements.

Discussion

This is an interesting exercise because we know that a connection is only created when your straight arm is raised up in the air (above the horizontal line) and the fibers of the pectoralis minor are positioned vertically. This position is used to stretch the fibers of the pectoralis minor.

Having examined the Deep Front Arm Lines (on page 54), we know that there are fascial connections between the pectoralis minor and the short head of the biceps, and the coracobrachialis at the coracoid process, but the functional connection (according to Thomas Myers' myofascial meridians) to the myofascial line is guaranteed when the arm is above the horizontal line or up in the air. We exploit this connection to stretch the DFL of the arm, with particular focus on the pectoralis minor and strengthening the lower trapezius antagonist. When performing the exercise, it is important to keep your spine in a neutral position and not to arch your thoracic or lumbar spine because this would change the position of your rib cage and cancel out the stretch.

Reduce the Level of Difficulty

Perform the exercise using your body weight or a resistance band.

ENERGY: MOVEMENT AS ELASTIC ENERGY

7

ENERGY

If you want to be quick, your fascia must be well trained!

7.1. ENERGY STRATEGY

The concept behind the ENERGY strategy is storing and then releasing energy.

As in the previous chapter, we will discuss in detail why and how I decided to include the ENERGY strategy into my training program.

Definition of Elastic Energy

Elastic energy is the potential energy associated with the elastic deformation of a solid or liquid (fascia exhibits both solid and liquid properties). Once the impulse of the elastic thrust has occurred, it turns into kinetic (movement) energy. Elastic recoil is the slingshot effect of the fascia.

What Does Deformation Mean?

Deformation is a change in the shape and size of an object caused by an applied force. Some materials exhibit viscous and elastic properties during deformation.

Let's look at how the fascial system responds to the application of force, as shown in a study by the English physicist, biologist, geologist, and architect Robert Hooke. Hooke was one of the great scientific minds of the 17th century and a key figure in the scientific revolution, giving his name to a number of physical laws [71].

In inorganic materials, response to applied force is governed by Hooke's law, which states that the deformation and its direction are proportional to the applied force. The response of human tissue is more complex, and there are four different types of deformation:

1) Pre-elasticity
2) Elasticity
3) Plasticity
4) Viscoelasticity

7.1.1. Pre-Elasticity

Pre-elasticity is an opposing movement.

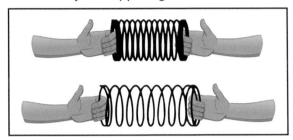

This concept can best be understood by looking at how a spring moves. Imagine that you are holding one end of a spring in either hand. As soon as you pull them apart, applying traction, the spring deforms and stretches, and its undulations or crimps diminish, storing energy. Simply put, the spring moves from being at rest to being tense.

Taking our fascial network as an example, it is as if the undulations or crimps are stretched: The resistance of the connective tissue is minimal.

The duration of the pre-elasticity phase depends on the degree of crimping of the collagen fibers and therefore differs depending on whether we are talking about the skin, muscles, tendons, or ligaments.

7.1.2. Elasticity

Elasticity describes the application of a force on a liquid or solid and represents a change of length. The liquid or solid returns to its original length when the force is removed.

Let's take the image of the spring again. If we continue to pull, linear deformation of the spring occurs by which the response is proportional to the force applied, and the stretch depends on the pull that we generate. If we stop applying force with our hands (pulling), the spring returns to its initial state.

The same thing happens with our fascial tissue.

7.1.3. Plasticity

Plasticity, or plastic deformation, describes the characteristics of substances to irreversibly deform after exceeding a limit in response to the forces applied.

7.1.4. Viscoelasticity

Viscoelasticity is defined as the elasticity of the fascial tissue fibers, together with the viscosity of the ground substance.

Viscoelasticity refers to a partially elastic, sometimes viscous material behavior. Viscoelastic substances therefore exhibit both liquid and solid properties. The effect depends on both time and temperature. The greater the viscosity, the thicker (less fluid) the liquid; the lower the viscosity, the thinner (more fluid) the liquid. Therefore, it could flow more quickly in the same conditions.

When a force is applied to a viscous substance, its shape permanently changes. In contrast, an elastic substance changes in length when a certain force is applied but returns to its original length when the force is removed.

The elasticity of our bodies is primarily caused by two structural proteins:

1) Collagen, which is slightly more solid and acts like a recoil spring
2) Elastin, which distributes elastic, partly plastic energy

Both are important components of fascial tissue and are each dependent on the other.

Taking the image of the spring again, the similarities end here. Compared to the spring, the excellent viscoelastic properties of fascia mean that the fascia can house fibers, some of which deform temporarily and others that deform permanently. A certain degree of deformation is irreversible even in the elastic phase: When a consistent tensile force is applied over a prolonged period, the tissue continues to deform proportional to time and no longer proportional just to the load. To put it simply, the balance between elasticity and viscosity changes.

It is interesting to note that once the load or force has been removed, it does not return immediately to its initial state. It therefore follows that, if a tissue is stretched multiple times, the final stretch is greater each time.

7.2. APPLICATION TO TRAINING

Having said all that, it is now time to apply it to our fascia training.

Are you ready to turn back the clock? This chapter is all about reliving our childhood as we have fun springing, bouncing, swinging, jumping, and much more. These movements stimulate the storage of elastic energy in our fascial tissue, which is fundamental for movement. This is generally true for all fascial structures surrounding the muscles, but it is particularly applicable to tendons. A healthy myofascial network works in a similar way to elastic springs that store a lot of kinetic energy (movement energy), and this function is essential for quick, powerful, and spontaneous movements like springing, bouncing, running, throwing, and dodging.

The human body is a masterpiece of "efficient collaborators", whereby the muscles and fascia divide and share the work required to perform any given movement.

Muscle work = The fascia relaxes = High energy consumption
Fascial work = The muscles relax = Low energy consumption

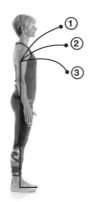

For a person with a healthy myofascial network, muscle work decreases as the upper body tilts further forward, increasing the tension of the fascial network.

1) Tilt of 20 to 30 degrees: predominantly muscular
2) From 30 degrees: fascial tension increases
3) From 90 degrees: predominantly fascial tension

7.3. SLINGSHOT EFFECT

I am sure that you must have used a slingshot when you were younger or at least seen somebody else use one.

How does it work? One hand holds the slingshot while the other holds the projectile and simultaneously pulls the elastic back so that it is tight. You take aim and then let go.

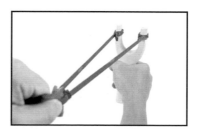

The tighter the elastic is pulled (stretching it), the faster and more powerful the projectile will be. Which part is the most difficult? That's right! Pulling the elastic back and not launching the projectile.

Our fascial tissue works exactly like this. This is the ENERGY strategy at work, and it is based on this effect of storing and releasing energy. This is the slingshot effect.

Schleip's contribution to this topic is very interesting: "Australian kangaroos can jump 9 to 13 meters (29 to 43 feet); this extraordinary jumping ability cannot be explained simply by the force of the muscle contractions of the lower limbs" [56].

By analyzing the jumping motion, scientists discovered the so-called slingshot effect for the first time. Think of the tendons and fascia of the legs of kangaroos as springs (elastic bands) that are in a state of pre-tension or pre-stretch. The subsequent conversion of elastic energy into mechanical energy enables them to jump incredible distances.

Modern portable ultrasound equipment was used to analyze the same division of work between muscle and fascia in human movements. It was discovered that

- the kinetic energy stored in the leg fascia of a human being is no less than that stored in a kangaroo and, in some cases, exceeds it (in this case the difference can be explained by the lever on which this elastic energy is released; look at the difference between a human foot and a kangaroo's foot), and
- we produce a significant amount of kinetic energy, in the manner described above, not just when we jump or run but also when we walk or throw something [57].

Let's see how it works in our body. Imagine throwing a tennis ball. First you load your arm and then you throw the ball.

This is known as pre-tension or pre-stretch: The fascial tissue is quickly pre-stretched, storing elastic energy, which is then released to perform the throwing movement.

If we look at our movements throughout the course of the day, we will find that we use the pre-loading strategy and slingshot effect to perform many of our actions: walking, running, jumping, throwing, pushing, punching, hitting—in other words, any action that requires accelerated movement.

Pre-tension is an opposing movement. This is generally true for all fascial structures surrounding the muscles (epimysium and perimysium that wrap around the muscle and tendons), but it is particularly applicable to tendons. Through this opposing movement, tension on the tendons and ligaments increases and energy is stored. Imagine stretching a rubber band. When you let go of the elastic band, the built-up energy is explosively released.

This strategy probably explains the surprising and incredible jumps performed both by athletes and normal people, even without particularly strong muscles. However, this phenomenon is not just reserved for sports activities. For example, when we walk we develop kinetic energy through the dynamic elasticity of the fascia of the Achilles tendon.

Through targeted and continuous training, your fascial tissue will become a high-performance aid to your muscles.

To clarify, a series of springs are required before the actual slingshot effect. In the springs, the elastic structures of the fascia are stretched for a short time, and elastic recoil is used. The slingshot effect can be thought of as the accumulation of these springs, whereby the energy stored in the myofascial structures is actively released.

Why Are Exercises That Use Elastic Energy Important?

They make the work easier! Why put in so much effort if you don't have to? The more you employ this strategy, the more energy your muscles will save because the fascia takes over most of the work. This energy saving has a positive effect on the whole body, while the exercises make the body more flexible.

The body is able to stretch, return to its desired shape, and remain elastic. Advantages of this strategy include the following results:

- It improves and supports our daily lives.
- It improves and supports our athletic performance.
- It makes us elastic.
- It makes the work easier.
- It makes the fascial tissue elastic and resilient.

7.4. ENERGY TECHNIQUES

I would like to begin by clarifying some of the terms I use to better explain the intensity and goals.

Springing: This action occurs in the lower limbs. They are characterized by one foot that stays on the floor for a low impact; there is no aerial phase. In relation to the whole body, the muscles are only involved through an increase in tension. The fascial structures, muscle fascia, tendons, and ligaments oscillate and bounce.

Swinging: The movements of the upper or lower limbs become wider. There are two types:

- **Swinging with gravity**: If, for example, you lean forward and swing your arms toward the floor, few elastic components of the fascial tissue are required.
- **Swinging against gravity**: Let's take the example of swinging from low to high, or vice versa. Swinging from high to low requires the involvement of the elastic structures when the movement is reversed to return to a standing position.

Bouncing: This is characterized by hopping with both feet or on one foot in an alternating and continuous rhythm, either on the spot or in motion. The aerial phase is much shorter than jumping, and the proportion of muscles involved is lower. In healthy myofascial tissue, bouncing increases the involvement of the epimysium and perimysium that surround the muscle and tendons, and increases the elasticity of the connective tissue.

Jumping: Both feet leave the ground, and there is a genuine aerial phase. If you jump with the fascial tissue pre-stretched (plyometrics), you will jump higher; the fascial structures adapt to the type of load. Jumping is always a combination of muscle contractions with support of the surrounding fascial tissue. The higher the muscle's demand for power, the greater the support of the surrounding tissue, and the greater the communication between the neighboring muscles, the better the jump will be. You first need to spring and bounce before you can jump. Jumping requires the harmonious interaction of the muscles and surrounding tissue.

To better understand this concept, it is helpful to compare functional training exercises with myofascial exercises. The most obvious differences can be seen in movements with weights and resistance bands and when jumping. In functional training, a resistance must be overcome, or maintained and stabilized, while functional fascial training uses flight, traction, and centrifugal force to effectively activate the fascial structures, exploiting the pre-stretch as much as possible.

Studies have shown differences in execution depending on the quality of the fascia. Jumps that involve significant movement of the knees and long contraction times on the floor are less reliant on fascia than jumps that involve less bending of the knees and short contraction times on the ground. In other words, the shorter and quicker the activation of the jump is, the greater the involvement of the fascial tissue.

Another important criterion is the quality of the stimuli. Our body overcomes resistance with concentric and eccentric movements. Combining the two represents the best training for our muscles, but rarely do our bodies undertake swinging movements while holding a full stretch, which is the best training for the fascia. We know that our body has to adapt to new demands (new movements). Because fibroblasts react to the quality of a stimulus rather than the quantity, the volume of fascia training will be less than traditional training.

To summarize, fibroblasts are not trained by the number of repetitions but by the quality of stimuli received. The quality of a stimulus depends on several fascial components. The more we integrate fascial components into a movement, the more effective it will be on the fascial structures. Let's take a look in greater detail.

Fascial Components in Swinging

1) Swinging

2) End to end swinging

3) End to end swinging with angle variations

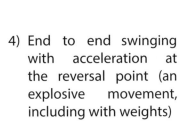

4) End to end swinging with acceleration at the reversal point (an explosive movement, including with weights)

Fascial Components in Bouncing and Jumping

1) Uniform bounces and jumps
2) Silent and uniform bounces and jumps, with significant changes to the angle of the knee
3) Stiff and uniform bounces and jumps, with little change to the angle of the knee
4) Jumping onto a platform and from the platform onto the floor; silent and uniform, with significant changes to the angle of the knee
5) Jumping onto a platform and from the platform onto the floor, with little change to the angle of the knee

Technique
- Stabilize your body.
- Pre-stretch a myofascial line.
- Let go: Use the stored energy to perform a controlled movement.

The greater the pre-stretch, the greater the slingshot effect. This effect is used and exploited in the starting blocks for short-distance sprinting.

Points to Remember:
- Although the fascial structures are very resistant to traction loads, they could be damaged if they are over-stretched or overloaded, hindering their ability to return to their original shape.
- Begin with small movements before increasing the range of motion.
- Listen to your body (there is always pre-tension).
- Always look for elastic recoil.
- Pick exercises that are suited to your current ability.
- Vary your movements.

7.5. ENERGY STRATEGY PLAN

We shall now look at the training plan or an appropriate inclusion of this plan in existing training. Please remember that in this chapter I am only talking about the ENERGY strategy and how to include it most effectively in a training context. In an actual training session I do include other FReE strategies, which you can find in the planning chapter.

ENERGY Exercises in a FReE Training Session
You will find ENERGY exercises in the following phases of a FReE training session:

- Warm-up
- Middle phase
- End phase

Repetitions: 3 to 10 per exercise
Training session duration: 30 to 55 seconds

ENERGY Exercises in a General Fitness/Sport-Specific Training Session
1. Warm-Up
- Start with light aerobic exercises to make sure that your body is sufficiently warm.
- Choose exercises for all the myofascial lines, with varying rhythms, speeds, and angles.
- Start with slow movements before increasing the speed but keep everything under control.

- Start with springs, swings, bounces, and then jumps.
- Include specific exercises for the sport you will play after the warm-up.

Repetitions: 3 to 10 per exercise
Warm-up duration: 5 to 20 minutes

Do some springing to release tension before a race and to be fully charged. Watch the great Usain Bolt before he takes his position at the starting blocks. Dance. Release all tension.

2. Middle Phase
- Include jumps at the beginning of the middle phase.
- Use springing to release tension during strength exercises.

3. End Phase
- Use springing and swinging.

ENERGY Exercises in a Stretching/Pilates Session
- Include swinging, springing, and pre-stretching in the warm-up. You will be surprised how much this improves your session.

Repetitions: 3 to 10

7.6. CATEGORIES AND COLLECTION OF EXERCISES

In this chapter the ENERGY exercises are categorized by level of difficulty and equipment:

- Categorized by difficulty: from simple to more difficult
- Categorized by equipment used: body weight – wall – equipment

7.6.1. ENERGY Body Weight Exercises

Flip Flops

Strategy
ENERGY – Elastic energy

Starting Position
Stand up straight.

Movement
Gently push your right heel down onto the floor. Then, with an energetic movement, lift just your heel (pushing your ankle forward), keeping your forefoot on the ground. Now push your heel back down onto the floor, pulling your Achilles tendon. Repeat 5 times then lift your leg and bend it at hip height, in the Stork position, a further 5 times.

Focus On
- Fixed point: forefoot; mobile point: heel.
- The push comes from the heel.

Work On
- Elastic energy of the SBL, specifically the plantar fascia and Achilles tendon.
- Joint mobility of the ankle.
- Stability and strengthening of the muscles of the LL (supporting leg).

Discussion
You may be asking yourself why I gave this exercise such a bizarre name. Flip flops acquired this name because of the strange noise they make as you walk. Pay attention the next time you wear them. When you walk, the back of the flip flops lift off the ground and hit your heels. This *flip flop* sound is the "music" that you should hear, or perhaps it is just that annoying noise of dragging sandals on the floor. For this exercise, imagine trying to walk noisily in a pair of flip flops by actively lifting the flip flops off the floor, causing the back of the shoes to hit your heels.

Change Stimulus
Perform the same movement but this time slightly move your back foot to increase the pull on the SBL; as you push your heel down onto the floor, raise your leg into Stork position.

VARIATIONS WITH LUNGE

Dynamic Lunge Starter

Take a bigger step backward to adopt a lunge position with your legs. Push your heel back, increasing the pull on your Achilles tendon, before returning energetically, but in a controlled manner, to the Stork position (use the energy created by pushing backward to bring you forward). As you move your right leg backward, lift your arms up to your sides at shoulder height. As you return to the Stork position, turn your upper body to the right and place your left hand on the outside of your right knee. Repeat 3 to 5 times and then change sides.

Supported Lunge

Perform the first part of the dynamic lunge exercise but this time place your hands on the floor or on a platform. Push off with your heel to catapult yourself with your body weight on your hands and raise your right leg up in the air. Repeat another 3 times and then change sides.

Supported Lunge With Bow

As you place your right forefoot on the floor, straighten your left leg and move your left hip toward the ceiling, increasing the stretch of the back of your left thigh. Thrust your right leg into the air, transferring your body weight onto your hands. Let yourself drop onto your left forefoot, cushioning the impact and stretching your left leg at the same time. Repeat the whole exercise with fluid and elastic movements.

Unsupported Lunge With Bow

Perform the same movement as described above but this time lift your hands off the floor as you bow to then catapult yourself forward; it is more destabilizing.

Combined Lunge

Perform the supported lunge exercise again but this time change the return, moving to a One Leg Deadlift position.

Combined Arm Lunge

Perform the backward lunge and lift your right arm above your head. As you move to the One Leg Deadlift position, lower your right arm toward the floor and stretch your left arm out in front of you. Imagine that you are holding a tennis ball in your right hand and that you want to throw it forward.

Elastic Lunge

Stay in the lunge position and bounce 3 times to raise and lower your pelvis. Push your heel back, return to the Stork position and switch legs. For greater stability, lift your arms out to your sides at shoulder height.

Strategy
ENERGY – Elastic energy

Starting Position

Stand up straight with your spine in a neutral position and your arms down by your sides. Before bouncing, walk on the spot for a minute, pushing your heels down onto the floor. Then start to raise and lower both your heels off the floor at the same time. The push must always come from your heels.

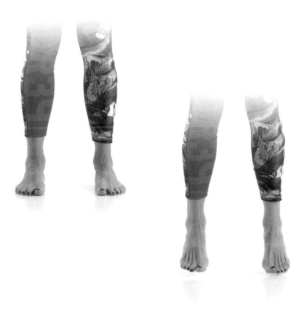

Discussion

Your landing should be silent. Land on your forefoot first and then your heel; let the shock absorbers of your foot (the springboard of your forefoot) absorb the impact. A silent and cushioned landing will not load the passive structures (joints) and will allow you to benefit from the slingshot effect (using the fascia more than the muscles) to make the movement ergonomic and functional. A noisy landing is often caused by landing on flat feet or heel first. A correct landing also stimulates lymphatic venous return upward. If performed correctly with the appropriate number of repetitions, and included in a training regime, these bounces could achieve optimal results against cellulite.

Movement

Start to bounce on the spot. Lift your feet slightly off the floor and land silently. Bounce 5 to 8 times, walk on the spot, and raise and lower your heels as described in the initial exercise. Then start to bounce again. Repeat the whole movement 3 to 5 times.

Focus On

- Pushing with your heels.
- Landing silently.
- Active kite.

Work On

- Elastic energy of the foot, ankle, and calf muscle (SBL).
- Improved circulation (also useful for anyone with cellulite).

Change Stimulus

- Bounce back and forth, sideways, and by twisting (moving your heels outward and inward). Does this remind you of anything? These are exercises used to warm up before skiing. Fantastic!

- Bounce with both your feet turned inward or outward.

- Alternate your legs: Bounce back and forth, and left and right on your right leg 3 to 6 times, then switch legs.

Test

Why take the time to do such a simple exercise? This test may give you the answer.

    With your legs straight, bend your upper body forward and try to touch the floor with your hands. Make a mental note of how far you are able to go and the tension you feel in your calf muscles, hamstrings, spinal erectors, back of your neck (nape), and head. Return to the starting position, perform the Elastic Bounces exercise, and then repeat the test. Most of you should see a net improvement in the stretch and tension of the SBL.

- With this variation we are taking a step back in time to my childhood. Hopscotch—does it remind you of something? I used to have great fun playing it! It's a real shame that we no longer see children playing hopscotch or skipping. We used to be a lot more "fascial" back in the day. This is how it works: Take two jumps forward on your right leg, sink into a half squat, jump forward on your right leg, and pick up the stone or ball off the floor. Turn around and repeat on your left leg. Repeat the whole movement 2 or 3 times. Vary the rhythm and the speed.

Strategy
ENERGY – Elastic energy

Starting Position

To get the most out of your training, you should choose an adjustable skipping rope so that you can adjust the length accordingly. To find the correct length for you, simply stand on the middle of the rope with one foot and lift the handles up on either side of your body. For beginners, the handles should be at shoulder height; for experts, at navel height.

Discussion

Ufff… how tiring I rediscovered skipping after many years, and I have to say that the first few times I tried it I found it very difficult to coordinate everything and move fluidly. I include skipping in many of my programs as a cardiovascular and fascial warm-up. I include it in Hamazon training, functional training in the training room, as a cardiovascular station, or during the day when I need to release some tension. As soon as I pull out the skipping rope, I hear my students say: "I can't do it... I don't know how to". Follow my three simple steps and you will start to improve in no time at all. The results are amazing, I promise you. If you ever get the chance to watch a boxer, you will see that he or she is balanced and full of energy without apparent effort. Why do you think that is? That's right! When you no longer find it exhausting, it is because you are exploiting the elastic energy. Think of a spring that is compressed before it releases energy. Work with your whole body, not with individual muscles.

Change Stimulus

Change the rhythm, skip on both legs, skip on one leg, swing the skipping rope twice but only jump once, skip forward or sideways, jump with your legs crossed... Have fun.

Movement

If you are a beginner and uncertain, try these steps.
Step 1: Bounce on the spot without the rope, simulating skipping. The important thing is to find a regular rhythm.

Step 2: Add the rope. Hold both handles in your right hand, bounce as before and add the movement of the skipping rope. The rope will not hamper your movement, but it will help you to synchronize your jumps with the rope.

Step 3: Put your feet in front of the skipping rope. Use your hands and wrists to swing the rope over your head. Keep your arms still and restrict the movement of your wrists. Jump just as the rope arrives back at your feet. Concentrate on the movement of your ankles. Bending your knees will make it even more strenuous. After two or

three jumps, relax by walking on the spot and then repeat. Gradually increase the number of jumps and reduce the number of breaks as you go. You should find that jumping gets easier after each break. It is better to skip in several short bursts rather than in just one long burst.

Focus On

- Using your ankles like two springs.
- Trying to jump silently. You should feel like a gazelle rather than an elephant.
- The less effort required, the better the myofascial response will be.

Work On

- Elastic energy.
- Improving agility.
- Improving coordination.
- Improving aerobic capacity.
- Strengthening of the muscles of the lower and upper body.

Standing Swing From Low to High

Strategy

ENERGY – Elastic energy

Starting Position

Stand up straight with your feet shoulder-width apart and lean your upper body forward, letting your head drop toward the floor.

Focus On

- Fixed point: feet; mobile point: upper body.
- Fluid, explosive, and controlled movements, without excessive effort or pulled muscles.
- Maintain the expansion throughout.
- Active kite.

Work On

- Elastic energy.
- Exploit the pre-stretch for elastic recoil.

Discussion

In this first swinging exercise, we take advantage of the expansion of the starting position on the SBL to release energy as we go up, only as far as an erect standing position or slightly further.

I urge you not to skip any of the stages; otherwise, you may end up with back pain. Start with simple swings and proceed as follows:

1. Swings
2. End to end swings
3. End to end swings with angle variations
4. End to end swings with acceleration at the reversal point (an explosive movement, including with weights)

Reduce the Level of Difficulty

When leaning forward, bend your knees slightly and keep your spine in a neutral position.

Movement

Gently swing your upper body forward and backward. Breathe out and curve your back by moving your arms between your legs, thereby increasing the pull on the SBL. Breathe in and lift your chest slightly as you move your arms slightly forward. Continue swinging your chest back and forth another three times, gradually increasing the range of motion each time. On the fourth swing, take a deep breath; push your heels down onto the floor; and, with an explosive but controlled movement, lift your chest and arms up, stopping in an upright position. Imagine throwing a ball behind you, expanding the Superficial Front Line. Breathe out slowly and lean forward again to return to the starting position. Repeat as desired.

Change Stimulus

- Swing from high to low: Perform the movement in reverse, starting from the extension and pre-stretching the SFL.
- Non-stop end to end swing with extension: Lean forward, as described in the previous exercise, and pre-stretch the SBL. From this position use the slingshot effect to lift you, but in the end to end swing increase the extension by pushing your pelvis forward slightly and making an arch shape with your body (also

see page 255, "Increase the Level of Difficulty"). Without stopping at the reversal point, use elastic recoil to return to the starting position and immediately perform again. Movements should be fluid, elastic, and controlled. It is a synergistic action between the elastic components (fascia) and the muscle.

- End to end sideways swing with change of trajectory: Perform the same movement as previously described, but when you are extended (upright), turn your upper body slightly to the left, thereby changing the trajectory.

- Start from a neutral, standing position. As you bend forward, move your upper body to the right, bringing your arms to the outside of your right leg. Use the pre-stretch of the SBL/LL to bounce back to the starting position. Repeat as often as desired and then change sides.

Increase the Level of Difficulty

- Accelerated and decelerated swing: At the reversal point, accelerate to go back down and come back up. This is a variation for more advanced students. Before attempting this and the subsequent steps, ensure that you have successfully completed the previous exercises.
- For fit people and athletes: Perform all the exercises with wrist weights. It is a great exercise for athletes because it involves accelerations, decelerations, and rotations that can be found in so many sports, like volleyball, basketball, golf, tennis, etc.
- Sideways swing in walking position: Take a step forward with your left leg and, with your arms raised above your head, stretch to create a slight extension, increasing the pre-stretch of the SFL. With a fluid but energetic movement, let your upper body fall forward, bringing your hands to the outside of your front foot. Exploit the increased pre-stretch of the SBL (slingshot effect) to return to the starting position.

- Sideways swing in walking position with rotation: As you return to the starting position, turn your upper body to the right, increasing the work of the LL/SL.

- For advanced students: Perform all these swings with weights.

Sideways Monkey

Starting Position

Bend your upper body forward with your feet shoulder-width apart, your arms straight, and your hands on the floor to the right.

Focus On

- Active kite.
- Explosive movements of your pelvis and legs.
- Be as light as a feather.
- Silent landing.

Work On

- Elastic energy of the legs.
- Proprioception.
- Balance with your weight on your hands.
- Coordination between movement and stability.
- Strengthening the arm and shoulder muscles.

Discussion

Monkeys are primates. They jump from branch to branch and from tree to tree and walk on all fours when on the ground. Because they have short trunks, their upper limbs are not the same length as their lower limbs. The way many so-called asymmetrical species move is striking: The absolute strength of their upper limbs is impressive given their body weight.

I chose sideways movement first because it is simpler than Front Monkey, an exercise where it is easier to lose control and fall onto your back.

Movement

Bend your knees slightly to load them. Transfer your body weight to your hands as you jump up with your pelvis and legs, landing silently to the right. Bend your knees again to reload the jump. Jump to the other side with an explosive and fluid movement of your pelvis and legs. Repeat as often as desired.

Change Stimulus

Perform the same movement but make it more fluid. Once you have landed on the right, lift your hands off the floor, move them to the left side, put them on the ground, and raise your pelvis explosively, dragging your legs to the left. Perform a fluid movement.

Increase the Level of Difficulty

Raise your pelvis until your upper body is perpendicular to the floor.

Strategy
ENERGY – Elastic energy

Starting Position

Adopt the position of a dog poised for attack (on all fours with your knees off the floor), with your pelvis moved toward your heels.

Movement

Move in a springing motion by lowering your pelvis toward your heels and your knees toward the floor and then lifting slightly upward. You should feel like your body is loaded, ready to lift your pelvis explosively into the air. With your fourth spring, transfer your weight to your hands and, with an explosive but controlled movement, lift your pelvis up. Land silently. Repeat 4 to 6 times and gradually increase the range of motion.

Focus On

- Active shoulder blades.
- Explosive, fluid, and controlled movements.
- Silent landing.

Work On

- Elastic energy of the legs.
- Proprioception.
- Balance with your weight on your hands.
- Coordination between movement and stability.
- Strengthening the arm and shoulder muscles.

Discussion

The perception of our body is literally turned upside down. This is the preparation for a handstand.

Activate the Arm Lines, push your pinkie down onto the floor using the outside of your hand to activate the Deep Back Arm Line, then push your thumb down gently to activate the Deep Front Arm Line. This creates a balance between the two lines. But we're still not finished. Now imagine that there are suction cups under your hands. Try to pull your hands gently toward your knees, and you will feel the pectoralis major and latissimus dorsi activating. Can you not feel it? Make sure that your shoulder blades are in active kite (lowered).

Reduce the Level of Difficulty

- Take smaller jumps.
- Perform the exercise against the wall as shown on page 251.

Watch out for jumps that are uncontrolled or too high: You may end up landing on your back. If you want to feel safer, perform the exercise next to a wall so that the wall will stop you from falling if you lose control.

Flying Attacking Scorpion

Starting Position

Stand up straight, with your feet hip-width apart. Keeping your legs straight, lean forward with your upper body until your hands are touching the floor. Take three steps forward with your hands and place them shoulder-width apart, with your fingers spread and your shoulder blades pushed down. Push your heels down onto the floor as much as possible. Your body should now be in the Camel position.

Focus On

- Maintain the activation of the Arm Lines.
- Active kite.
- Fluid and controlled movements.
- Back leg straight.
- Silent landing.

Work On

- Elastic energy of the legs.
- Proprioception.
- Balance with your weight on your hands.
- Coordination between movement and stability.
- Strengthening the arm and shoulder muscles.

Discussion
See page 243.

Movement

Bend your right knee, lifting your foot off the floor, and bring it toward your upper body. At the same time, slightly bend your left knee. Prepare to kick with your right leg up toward the sky. Moving your knee toward your chest creates the energy you need to energetically extend your leg upward in a controlled motion. Repeat another two times.

On your third repetition, push off more strongly with your pelvis to lift your left foot off the floor. Repeat another 2 times and then change sides.

Strategy

ENERGY – Elastic energy

Starting Position

Get on your knees, with your feet in hammer position.

Movement

Raise your arms slightly as you let your body fall to the right.

As you land on the floor, ensure your hands are slightly more than shoulder-width apart (see Gecko Push-Up Against the Wall, page 250). Bend your elbows to cushion the load born by your hands on the floor.

Straighten your arms explosively to return to the starting position.

Without pausing, let your body fall to your left and repeat all the points described above.

Focus On

- Active kite.
- Active core.
- Harmonious, fluid, explosive, and silent movements.
- Distribution of force on the Arm Lines.

Work On

- Elastic energy of the arms.
- Strengthening the muscles of the arms and core.
- Coordination.
- Correct amount of force.
- Acceleration and deceleration.

Discussion

What difficulties might we face when doing a push-up? Difficulties caused by a lack of strength, incorrect activation, or a lack of coordination can be resolved with the appropriate progressive exercises, which, if used well, produce excellent results. But that's not enough. I want to focus on the joints, like the wrists and shoulders.

1. Focus on correct technique.
2. Focus on mobility and dynamic stretching preparation.

The second point is important to allow your joints to move freely and without restrictions. Without this freedom of movement the exercise cannot work. This relies on the instructor's ability to teach (correct technique and plan). Let me give you an example. Before performing any push-up or any exercise requiring you to put weight on your hands, do the mobilization exercises like Spider Woman (page 131), the activation of the Arm Lines (page 54), and so on. Get the idea? I go into more detail in my course planning.

Traditional strength training push-ups performed with narrow, bent elbows and straightening the arms primarily work the triceps, pectoralis major, rhomboid, and middle trapezius, which is fine from a muscle-strengthening perspective. We try to be different by providing another type of stimulation. As you perform the push-up (see page 245), keep your collarbones open and move your elbows slightly away from each other, creating space between your shoulder blades.

Now try and bring your elbows closer together: Push your shoulder blades toward your buttocks, maintaining the space between your shoulder blades. This strengthens but in expansion.

I'm not saying that one method is better than the other. The goals are not the same. Vary your exercises and stimulate in different ways: This is what makes you strong and ready to react to any situation.

Reduce the Level of Difficulty

Place your hands on a platform, like a step, when you fall to reduce the range of motion. Reduce the height of the platform gradually as you progress. See Gecko Push-Up Against the Wall on page 250.

Change Stimulus

Move a little in all directions to give different stimuli. The only thing I would say is that your shoulder blades must always be in the right place; otherwise you will load your shoulders.

Strategy

ENERGY – Elastic energy

Starting Position

Get on all fours with your toes tucked under. Move your pelvis toward your heels, keeping your hands on the floor and with your shoulder blades pushed toward your buttocks.

Movement

Imagine loading a slingshot to throw a stone. Push your feet into the floor and your heels back, loaded. Then, with an explosive but controlled movement, push your upper body up between your arms, extending your legs (lift your knees off the ground). From this plank position, bend your arms to lower your upper body toward the floor. Taking advantage of the slingshot effect of the arms, straighten them by pushing your pelvis toward your heels to return to the starting position like a spring.

Let's add another movement. After you have bent your arms (see photo above), straighten them explosively. Lift your hands off the floor during the short aerial phase before putting them back on the ground again. Go back down again quickly, cushioning your impact with the floor, then push off with your arms to return to the starting position.

Focus On

- Active kite.
- Neutral position of your spinal column.
- Elbows close to sides when descending.
- Active core.
- Fluid, explosive, and silent movements.
- Distribution of force on the Arm Lines.

Work On

- Elastic energy of the arms.
- Strengthening the muscles of the arms and core.
- Coordination.
- Correct amount of force.
- Acceleration and deceleration.

Discussion

If you are ready to perform the whole sequence, imagine moving like a cricket. Load your legs ready to explode upward, sideways, or forward but make sure that your movements are harmonious and the right intensity. To this end, it is important that you do not skip the motor and technical learning stages as you shall require a good deal of force distribution (fascia and muscle), as well as appropriate energy expenditure.

I could dedicate a whole chapter to push-ups alone, but I'll stop myself and just detail the key points. The most important thing is that your shoulder blades are always activated (active kite), particularly when pushing up; otherwise you will end up with shoulders that are internally rotated (greater stress on the scapulohumeral joint and excessive work on the upper trapezius). Activating the shoulder blades will help to activate the latissimus dorsi. Don't forget the important points of the plank (see page 63 onward).

I'll just give you one piece of advice: Read the book *Allenamento Funzionale. Manuale Scientifico* (The Functional Training Bible) by Guido Bruscia, which contains several variations and lots more besides.

Reduce the Level of Difficulty

- Use a stable platform (such as a step or Plyo Box) to put your hands on and reduce the range of motion. Gradually reduce the height of the platform.
- Only perform the forward and backward movements, keeping your knees on the floor.
- Perform the forward movement by bending your arms and resting on your knees.

- Push up on the ground: Place a sand bag lengthways between your arms. As you lower your upper body, this will act as a support, reducing the ROM.
- Only perform the first part of the starting position and move directly to the plank position and then back again. When you move forward, start the movement by pushing your forefoot onto the floor (fixed point). Meanwhile, push your heel back and your sternum (fifth rib) forward (opposing points) without changing the position of the kite. This will activate and stretch the SFL, which will help to distribute the force throughout your body. Perform these movements harmoniously but explosively. Repeat 5 to 10 times, concentrating on elasticity and on the distribution of the force, as well as activating the kite and the core.

Increase the Level of Difficulty

In the starting position, keep your knees raised off the floor. In other words, your knees never touch the floor.

Starting Position

Adopt a dog posture (on all fours with your knees raised).

Movement

Take small springs by lowering your knees slightly closer to the floor while at the same time bending your elbows toward your knees. Then return to the starting position. With your fourth spring, lift your hands and feet off the floor by explosively moving your pelvis upward. Cushion your landing by slightly bending your arms and knees. Repeat 4 times.

Focus On

- Active kite.
- Active core.
- Fluid and explosive movements.
- Silent landing.

Work On

- Elastic energy of the arms and legs.
- Strengthening of the muscles of the arms, legs, and core.
- Coordination.
- Correct amount of force.
- Acceleration and deceleration.

Discussion

It is not easy to pretend to be a grasshopper, and not everyone finds this exercise simple to do. It requires a great deal of synchronization and coordination between your arms and legs.

Reduce the Level of Difficulty

Jump and move with your arms only or just with your legs.

Increase the Level of Difficulty

Jump forward, to the side, or diagonally.

7.6.2. ENERGY Exercises Against the Wall

Gecko Push-Up Against the Wall

Strategy
ENERGY – Elastic energy

Starting Position

Stand up straight about 3 to 5 feet (1 to 1.5 meters) away from the wall, depending on your height (the taller you are, the further away from the wall you need to be). Keep your spine in a neutral position and activate your kite.

Focus On

- Active kite.
- As you move, keep your whole body aligned (like a plank of wood).
- Perform elastic movements and cushion the impacts without making any noise.

Work On

- Elastic energy of the arms.
- Strengthening the muscles of the core and of the scapulohumeral joint.

Increase the Level of Difficulty

Flying gecko against the wall: Fasten your seatbelts because we're about to take off! Perform the same exercise but this time as you fall forward, bend your arms just slightly and lift both your feet off the floor. Catapult yourself back by putting your feet on the floor (slightly bending your knees). Push off with your feet on the floor to immediately jump back toward the wall. Change the angle with each jump.

Movement

Lift your heels off the floor as you fall forward toward the wall. Stop yourself falling by placing your hands on the wall, with your right hand slightly higher than your left hand, and bending your elbows slightly so that your whole body slides forward.

With an explosive and controlled movement, push yourself backward and take your hands off the wall, returning to the starting position but keeping your heels off the floor. Fall toward the wall again but this time reverse the position of your hands (left hand higher than your right hand), and move your elbows slightly outward. Use the elastic energy to catapult you back to the starting position without lowering your heels to the floor. Repeat several times, changing the angle of your arms each time.

Discussion

These elastic bounces, which are performed fluidly, elastically, and without great effort in all directions, are a great alternative to the same exercise performed on the floor. Let's compare. The exercise against the wall can be done by almost anyone. In addition, it makes greater use of the elastic components (fascia) and less so the muscles. In contrast, the floor exercise is only suitable for people with sufficient muscular strength, elasticity, and coordination and who have already mastered the motor skills. What's more, the floor exercise is more reliant on the muscles and loads the joints if performed incorrectly.

Strategy

ENERGY   Elastic energy

Starting Position

Adopt a dog posture (on all fours with your knees raised), with your feet close to the wall. Transfer your weight to your hands as you lift your right foot off the floor and place your forefoot on the wall so that your right knee is even with your hips. Activate the kite and abdominals, and then lift your left foot off the floor and place it on the wall at the same height as your right foot.

Push forward slightly and you should feel the Back Arm Lines activating (particularly the lower part of the trapezius, which connects to the coccyx). Push your shoulder blades toward your buttocks. Imagine that you want to push the floor forward.

Movement

Activate your abdominals and your arms; spring twice; and on your third spring, lift your pelvis and feet upward with an explosive but controlled movement, jumping a little into the air. Repeat the jump but this time go lower. Vary the exercise: Jump twice upward and twice downward. Repeat 4 to 6 times and gradually increase the range of motion.

Focus On

- Opposing points: hands and shoulder blades.
- Supporting points of the hands: pinkie and thumb.
- Active kite.
- Fluid and controlled movements.

Work On

- Elastic energy of the legs.
- Proprioception.
- Balance with your weight on your hands.
- Coordination and stability.
- Strengthening the arm and shoulder muscles.

Reduce the Level of Difficulty

- Take smaller jumps.
- Perform the exercise on page 243.

Strategy
ENERGY – Elastic energy

Starting Position

Get on all fours with your hands shoulder-width apart. Raise your straight right leg to the side and place your foot on the wall at hip height. Your right heel should be in line with your left knee (draw a perpendicular line from your knee to the wall: Your right leg and heel should be parallel to this line), forming a rectangle.

Movement

Rhythmically bend and slightly straighten your right knee. On your third repetition, breathe in; with an elastic movement lift your foot off the wall and up into the air, with your leg straight. Return silently to the starting position. Repeat 4 to 6 times and then change sides.

Focus On

- Fixed point: knee on the floor; mobile point: foot against the wall.
- Active kite.
- Stay on all fours, with your heel in line with your knee.

Work On

- Elastic energy.
- Stretching the inner thigh (DFL).
- Functional mobility of the hip.
- Strengthening the muscles of the LL/FL/AL.
- Stabilizing the scapulohumeral joint.

Reduce the Level of Difficulty

- Perform the same movement but this time lower the leg that is against the wall below your pelvis. Lift your foot off the wall, raise your leg a bit higher, and put your foot back on the wall. Return to the starting height with your next jump.
- Place a platform under your knee on the floor.

Change Stimulus

Vary the position of your foot: a bit lower, a bit higher, turned slightly inward or outward.

Discussion

An interesting aspect of this exercise is the stretching and activation of the DFL. Let's find out why. Let me emphasize again that in this position you are going to stimulate your pelvic floor muscles through your adductors because they are closely connected (see DFL, fascia of the pelvic floor, levator ani). At the same time, there is also a connection to the adductors of the left leg.

If you start to slightly vary the position of your foot, as described in the "Change Stimulus" section, you will stimulate the adductors and muscles of the pelvic floor in different directions.

This logically leads to the following question: Is this exercise also a workout for the pelvic floor? And can it improve a prolapsed (dropped) bladder? Yes to both! At the end of the book you will find a list of exercises that can help to make the pelvic floor elastic, permeable, and resilient.

7.6.3. ENERGY Exercises With Equipment

Sideways Swing on Plyo Box

Strategy
ENERGY – Elastic energy

Starting Position

Stand up straight about 20 inches (50 cm) from the Plyo Box, with your feet hip-width apart. Place your right heel on the Plyo Box. Turn your upper body to the left, with your arms raised above your head. Push down and apply light pressure and traction with your heel on the Plyo Box to activate the lower part of the SBL (isometric pre-stretch).

Movement

With an energetic but controlled movement, let your upper body fall forward, turn your upper body slightly to the right, and position your straight arms on the outside of your right leg.

Without stopping at the reversal point, use elastic recoil to return to the starting position and immediately perform again. Control the movement deceleration as you lower your upper body. Repeat as often as desired and then change sides.

Focus On

- Fixed point: heel; mobile point: chest.
- Keep both legs straight.
- Explosive, fluid, and controlled movements without excessive effort or pulled muscles. Expand into your skin "wetsuit".
- Active kite.
- Accelerate and decelerate.

Work On

- Elastic energy of the SBL/LL/SL.
- Exploit the pre-stretch for elastic recoil.
- Proprioception and control.
- Acceleration and deceleration.
- Muscle strengthening.

Change Stimulus

Start by swinging from low to high. As your upper body swings down, move toward your heel, knee, or thigh, changing and varying the tension.

Discussion

Unlike the Sideways Swing movement described on page 241 (bottom left), because this exercise uses a platform you should feel increased stretching in the hamstrings and calf muscle region (back of the leg).

Reduce the Level of Difficulty

- Do not turn your upper body in the starting position.
- Reduce the height of the box.

Increase the Level of Difficulty

Use wrist weights.

Sideways Swing With Resistance Band

Starting Position

Use a medium- to high-level resistance band that is at least 4 feet (120 cm) long. Tie a knot at one end so that you can hold it with your hand. Put your right foot on the other end, cross the resistance band behind your body, and slide your left hand into the knot. Extend your left arm above your head, pulling the resistance band tight. Take a step back with your left foot.

Movement

With an explosive movement, turn and bend your upper body forward, bringing your left hand to the outside of your right foot. You should feel the tightness of the resistance band increase. In this position, try to expand against the resistance band, keeping it in contact with your sacrum, lumbar spine, thoracic spine, and head. It will be a piece of cake to come back up because you can use the pre-stretch of the resistance band to return to the starting position. Repeat as often as desired.

Focus On

- Fixed point: feet; mobile point: chest.
- Keep both legs straight.
- Fluid, explosive, and controlled movements, without excessive effort or pulled muscles.
- Expand against the tight resistance band.
- Active kite.

Work On

- Elastic energy of the SBL/LL/SL.
- Exploit the pre-stretch for elastic recoil.
- Proprioception and control.
- Acceleration and deceleration.
- Muscle strengthening.

Discussion

It is very important to stress that using the resistance band is vital to this exercise. It is easier to understand and feel the pre-stretch strategy with the resistance band than just with your own body weight. If the resistance band shifts, it is because you have lost the expansion: You get compressed by the force of the resistance band, and you lose the tension.

Strategy
ENERGY – Elastic energy

Starting Position and Movement

Start from the extension to pre-stretch the SFL. Let your arms fall as you push your upper body forward. Continue the movement of your arms, bringing the clubs behind your back. Bend your knees slightly. At the reversal point, use elastic recoil to return to the starting position and immediately perform again.

Discussion

I won't dwell on all the details, which you can find in the "Discussion" section on page 240. The only thing that changes is the use of the wooden clubs, which are fantastic because they are a weight that facilitates the swing, extends the lever, and enhances the effects of the exercise. Wooden clubs are not a new piece of equipment and are used by artistic gymnasts and other disciplines. I have also found them to be an invaluable tool in fascia training, since they are lightweight and easy to use.

Increase the Level of Difficulty

- Start from extension to pre-stretch the SFL. Bend your upper body forward as far as it will go; you should feel the increased pre-stretch of the SBL. From this position, take advantage of the slingshot effect to lift yourself back up, increasing the extension and pushing your pelvis slightly forward so that your body is arch-shaped. Without stopping, go back down again.

- Challenge your sense of balance in an energetic, elastic, and controlled manner. Perform the entire movement as already described but on just one leg, combining the Stork position in expansion and the One Leg Deadlift. Repeat the entire sequence 5 times with your right leg raised (without ever lowering it to the floor) and then change sides.

Lunge With Horizontal Swing With Wooden Clubs

Strategy
ENERGY – Elastic energy

Starting Position

Stand up straight with a wooden club in each hand.

Movement

Raise your arms to shoulder height and move them to the right. At the same time, take a big step forward with your right leg and bend your knees. At the reversal point, use elastic recoil to accelerate as your arms move to the left. At the same time, take a big step back with your right leg. Perform this exercise rhythmically another 5 to 10 times, taking advantage of the acceleration and deceleration of your arms and only moving your right leg back and forth. Then repeat from the beginning with your left leg.

Focus On

- Fixed point: left foot; mobile points: chest and right foot.
- Dynamic, fluid, and controlled movements, without excessive effort or pulled muscles.
- Active kite.
- Accelerate and decelerate.

Work On

- Elastic energy of the LL/SL.
- Exploit the pre-stretch for elastic recoil.
- Proprioception and control.
- Acceleration and deceleration.
- Muscle strengthening.

Discussion

Don't just move your arms with the clubs from left to right! To do so would only tire your shoulders, and you would not benefit from the elastic energy of your upper body (think of your upper body like a wet T-shirt: Wring it out to rotate and release energy to return).

Change Stimulus

- As your arms move to the right, take one step forward with your right leg. As your arms move to the left, use the elastic recoil to take a step forward with your left leg. Keep moving forward like a walking lunge. Accelerate and decelerate with fluid and energetic movements.
- Change the angle.
- Use wrist weights.

Strategy
ENERGY – Elastic energy

Starting Position

Pick up a weight in your right hand, take a step forward with your left leg, and raise your right arm straight above your head. Stretch and create slight extension, increasing the pre-stretch of the SFL.

Movement

With an energetic but fluid movement, let your upper body fall forward, bringing your right hand to the outside of your left leg. Exploit the increased pre-stretch of the SBL to return to the starting position. Take advantage of the reversal point to accelerate. Repeat as often as desired.

Focus On

- Fixed points: feet; mobile point: chest.
- Keep both legs straight.
- Fluid, explosive, and controlled movements, without excessive effort or pulled muscles.
- Expand into your skin "wetsuit".
- Active kite.
- Accelerate and decelerate.

Work On

- Elastic energy of the SBL/LL/SL.
- Exploit the pre-stretch for elastic recoil.
- Proprioception and control.
- Acceleration and deceleration.
- Muscle strengthening.

Discussion

In this exercise, everything that you have learned and practiced with the exercises Sideways Swing in Walking Position (page 241) and Sideways Swing With Resistance Band (page 254) will come in handy. The Swing With Dumbbells exercise is only suitable for advanced students because you need to know how to create the pre-stretch and expansion that will enable you to take advantage of the recoil; otherwise you risk hurting your back.

It is a very useful warm-up exercise for many athletes. How many sports can you think of in which this movement might be performed? Tennis, volleyball, basketball, and others come to mind. This is a pre-stretch exercise for the whole body.

Increase the Level of Difficulty

Pick up a weight in your right hand, take a step forward with your right foot, and raise your right arm straight above your head. Pre-stretch the SFL. Explosively turn and bend your upper body and right arm to the left as you take a step forward with your left leg and bend both your knees. Return to the starting position. Change the rhythm: accelerate and decelerate.

Reduce the Level of Difficulty

Perform the exercise Sideways Swing in Walking Position (page 241), which is more elastic.

Strategy
ENERGY – Elastic energy

Starting Position

Stand up straight, pick up one soft kettlebell in each hand, and adopt the kite position.

Movement

With a rhythmic and elastic movement, slightly turn your upper body, and move your right arm forward and your left arm backward. Continue to swing fluidly and accompany the rhythm of your arms with small flexions and extensions of your knees. Gradually increase the rotation of your upper body and the range of motion of your arms. Keep your head and gaze facing forward. Repeat as often as desired.

Focus On

- Fixed point: head; mobile points: upper body and pelvis.
- Rhythmic, fluid, and controlled movements, without excessive effort or pulled muscles.
- Active kite.
- Accelerate and decelerate.

Work On

- Elastic energy of the LL/SL.
- Functional mobility of the thoracic spine and shoulder joint.
- Exploit the pre-stretch for elastic recoil.
- Coordination between your arms and legs and trajectory of the kettlebells.
- Muscle strengthening.
- Acceleration and deceleration.

Discussion

This is a very functional movement because it reproduces the movement of the trunk when walking and at the same time improves breathing.

Reduce the Level of Difficulty

Perform the exercise without weights, with lighter weights, or with wooden clubs.

Increase the Level of Difficulty

- As you continue to swing, take a step forward with your left leg as your right arm goes forward. Hold that position and continue to swing. After three swings, take a step forward with your right leg as your left arm goes forward.
- Perform one swing with each step.

Synchronized Swinging Walking Lunge With Soft Kettlebells

Starting Position

Stand up straight, pick up one kettlebell in each hand, and ensure your back and pelvis are in a neutral position.

Focus On

- Fixed point: head; mobile points: upper body and pelvis.
- Rhythmic, fluid, and controlled movements, without excessive effort or pulled muscles.
- Active kite.
- Acceleration and deceleration.

Work On

- Elastic energy of the multiple lines involved.
- Functional mobility of the thoracic spine and shoulder joint.
- Exploit the pre-stretch for elastic recoil.
- Acceleration and deceleration.
- Coordination between your arms and legs and trajectory of the kettlebells.
- Muscle strengthening.

Discussion

The difficulty of this exercise lies in coordinating your body's movements with the swinging of the kettlebells.

The key to this exercise, with the kettlebells moving in the same direction, is timing because the loading is used to then explode forward in a measured and controlled manner, giving the movements fluidity and harmony, like in a dance. This is what I mean by timing: Imagine you are dancing the Fox Trot, Jive, Salsa, or Tango, and your partner is doing exactly the opposite to what you are doing. If your timing is not synchronized, the dance becomes tiring both mentally and physically.

Movement

Start to swing your arms back and forth, with the kettlebells moving in the same direction, at an appropriate rhythm. When both the kettlebells are behind your back and are about to be catapulted forward, use the recoil to take a step forward with your left leg.

Hold this position and continue to swing back and forth twice more. On the third swing, take a step forward with your right leg. Continue this walk for the desired number of repetitions.

Increase the Level of Difficulty

- Take a step forward with your left leg and continue to swing the kettlebells. On the third swing, while your arms are behind your back, perform a lunge (bend your right knee until it touches the floor in a half squat) and use the recoil of the kettlebells to arise from the lunge position and take a step forward with your right leg. Perform this sequence smoothly for the desired number of repetitions.
- To make the whole exercise more fluid, you could also swing with each movement. As you lunge, move the kettlebells behind your back then move them forward as you use the recoil to take a step.
- Remember: Use rhythmic, fluid, and controlled movements, without excessive effort or pulled muscles.

Infinite Swing With Kettlebell

Strategy
ENERGY – Elastic energy

Starting Position

Stand up straight, hold a kettlebell in your right hand, and take a step forward with your left leg.

Movement

Swing and externally rotate your right arm until it is behind your back. At the same time, transfer your body weight to your right leg. At the reversal point, slightly rotate your arm inward, moving it forward together with your body weight. Keep swinging, changing the trajectory and the speed, as your weight shifts to your left leg and your arm moves around to your left side and behind your back, keeping it rotated inward. Reverse the motion to bring the kettlebell back to your right side with your right leg supporting your body weight. Continue for the desired number of repetitions then switch sides.

Focus On

- Rhythmic, fluid, and controlled movements, without excessive effort or pulled muscles.
- Active kite.
- Acceleration and deceleration.

Work On

- Elastic energy of the multiple lines involved.
- Functional mobility of the thoracic spine and shoulder joint.
- Coordination between your arms and legs and trajectory of the kettlebells.
- Muscle strengthening.

RELEASE: FASCIAL RELEASE

RELEASE

Fascia loves to be squeezed, requires well-measured pressure, and likes to slide.

8.1. RELEASE STRATEGY

As in the previous chapter, we will discuss in detail why and how I decided to include the Myofascial RELEASE strategy into my training program. It's a simple technique with instant results and a wide range of stimulations for our central nervous system (CNS), including pressure, friction, shakes, vibrations, swinging, traction, and touch.

In my book *Pilates per lo Sport* [Pilates for Sport] [3], I talked about BBR, Body Ball Relaxing, dedicating a whole chapter to self-massage as part of your training program. Today it is more of an integral part of my training than ever—but with greater awareness.

Fascial tissue acts like a filter for all the cells in your body. It processes metabolic waste products until they can be transported by the blood or lymphatic system and also supplies the cells with fresh substances. If the fascial filter gets blocked, the exchange of both incoming and outgoing materials is inhibited. Cells that are not sufficiently supplied will die. We use the RELEASE strategy to stimulate the

261

fascia, thereby enhancing our metabolism and our general health. Balls and foam rollers are excellent tools for self-massage and are ideal for relaxing the myofascial structures and stimulating greater regeneration.

The RELEASE strategy works specifically on the expansive myofascial tissue by acting on the muscles and nerves. The pressure applied by a ball or foam roller on the affected area rehydrates the tissue and eliminates tension, leaving the muscles more elastic, energetic, stronger, and with their functions reactivated.

Think of a pond with little or no exchange of water: After a while it begins to stink and putrefaction sets in. In contrast, a lake with a continuous exchange of water stays clean and fresh.

The RELEASE strategy specifically targets the interstitial fluid (the lubricant of the muscles) to increase the fluidity of movement. Moreover, it releases the fascial tissue that covers the muscle from tension, stagnation, and stasis, enabling the muscle to slide freely in its sheath. Think of it as wringing out a sponge full of stagnant water so that it can soak up clean water. The effect of the RELEASE strategy is instant, particularly when it comes to sport. A well-lubricated muscle "motor" will yield the highest sporting performance and help prevent muscle injury (see "Water" in section 1.3.1). That is why myofascial massage is particularly recommended for athletes, although many people can benefit from its effects.

Objectives of the Myofascial RELEASE Strategy:
- It rehydrates and hydrates the tissue.
- It eliminates tension.
- It improves flexibility.
- It improves regeneration.
- It alleviates pain.

What Are the So-Called Trigger Points?

A healthy muscle is loose, elastic, and not painful to the touch. Think of an 8-inch (20-cm) piece of rope. If you tie a knot in the middle, the ends get shorter. This is what happens with muscle tissue or with its associated fascia: Trigger points are created in the neighboring tissue.

Myofascial trigger points, commonly known as trigger points, are hyperirritable knots in contracted bands of muscle, characterized by palpable nodules.

What happens when a trigger point forms? The contracted bands of muscle come together, get as close as possible, and lose their elasticity, while the other fascia (which are not involved) are forced to stretch, like a tight elastic band. All of this gives rise to a loss of both passive and active dynamic mobility because part of the band of muscle is already stretched which limits the size of the entire structure. Another consequence is reduced muscle strength, because not all the fibrils can be recruited during contraction, resulting in a risk of damage. This is an athlete's worst nightmare. The most common cause of this is probably the overloading or incorrect loading of the muscle. At the trigger point, the supply of oxygen and nutrients is inhibited, resulting in continuous contraction.

The most common causes are

- too much time in front of the computer,
- incorrect posture,
- sitting for too long in the car,
- stress,
- negative emotional states, and
- poor diet.

But trigger points can also be caused by an excessively sedentary lifestyle, in which many hours are spent in the same position without giving your muscles the chance to stretch and relax.

Types of Trigger Points

There are various types of trigger points, each with their own unique characteristics:

- Primary (or central) trigger points are the most typical, are found in the middle of the muscle belly, and are generally the easiest for patients to recognize and report.
- Secondary (or satellite) trigger points form around the primary trigger point, which remains the first to be treated.
- The trigger points at attachment points are found in the tendons.
- Diffuse trigger points affect another part of the body and are associated with postural deformities like scoliosis or hyperlordosis.
- Primary or secondary active trigger points give rise to referred pain and are painful to the touch.
- Latent (or inactive) trigger points do not cause referred pain and are not painful, but they lead to muscle stiffness and can be reactivated following stimulation.

8.2. CONTRAINDICATIONS

The RELEASE strategy is not indicated in the following cases:

- If you have stitches, fractures, or open wounds
- If you are taking anticoagulants
- If you have malignant tumors, osteo-porosis, acute rheumatoid arthritis, diabetes, hypersensitive skin, or diseases of the circulatory system (such as edema, hematoma, high or low blood pressure)

In addition, pay attention to the following areas:

- The coccyx (to be avoided due to risk of fracture)
- The rear of the 11th and 12th lower ribs (floating)
- The xiphoid process (lower end of the sternum)

- The abdominal area (the front of the body, from the pubic bone to the area below the ribs)
- The cervical vertebrae and the upper cervical region (arteries, cranial nerves, and spinal nerves)

8.3. EQUIPMENT

So many things can be used for self-massage, including balls of various sizes, foam rollers, sensory balls, and tennis and golf balls. My favorite are soft sensory balls because they stimulate the tissue below the surface. The extent to which the myofascial tissue may change using these balls depends on how much pressure is applied, for how long, and how quickly or gradually.

Foam rollers and balls measuring 5, 6, or 7 inches (13, 15, or 18 cm) in diameter work well for most people, both because they are three-dimensional and because the user can adjust their firmness by filling them with more air or letting some out.

Beginners, people taller than 6 foot 5 (1.98 m), and people weighing more than 90 kilograms should use the biggest ball (7-inch [18 cm] diameter). Anyone with extremely contracted muscles or muscles that are painful to the touch, or who has multiple trigger points, should also begin with a ball of this size. This ball is best used for a delicate and general response and release.

Use the 6-inch (15 cm) diameter ball if you do not experience any discomfort after using the 7-inch (18 cm) ball for self-massage. This ball can be used for various parts of the body, such as the quadriceps and latissimus dorsi, and is suitable for people who are more heavily built.

The smallest 5-inch (13 cm) ball is more intense and is recommended for a more advanced and specific training program. It can be used in more restricted areas, such as deep in the hip flexor muscles or in the rotator cuff.

Stop the self-massage if you experience vertigo or nausea. You should also be aware that some regions of the body are more prone to injury if excessive pressure is applied, which is why it is important to proceed with great caution and care when working close to or around these areas.

8.4. MYOFASCIAL RELEASE TECHNIQUES

Applying the correct amount of pressure reduces fascial and muscular tension. I would like to begin by clarifying some of the terms I use to better explain the intensity and goals.

Massage: On the trigger points – apply gentle pressure by tiny movements of no more than half an inch (1 cm) back and forth – eliminate tension.

Support: On trigger points – hold the position, press for 3 seconds and relax – eliminate tension.

Slide: On a broader area in the direction of the myofascial lines – apply gentle pressure and let your skin slide over the ball/foam roller – elastic effect.

Roll: On a broader area – roll slowly or quickly; if you roll slowly, move in the direction of your heart after each breath – squeezing effect.

Loosen up: On trigger points – stay still on a painful point, continue to breathe, and apply gentle pressure on the ball every time you breathe out, until the tension subsides, then move half an inch (1 cm) in the direction of your heart and repeat. On certain trigger points you could also move in small circles or draw a star.

This range of techniques on the fascia helps to soften them, thanks to a combination of mechanical, biochemical, and neurological responses that are still being studied.

Points to Remember:
- The goal is not to die of pain on the ball as per the no pain no gain philosophy since this would achieve the opposite effect.
- Never hold your breath. Breathe constantly throughout.
- If you find a painful point, reduce the pressure, try to stay in that area, breathe deeply, and make small movements of no more than half an inch (1 cm) until the pain abates.
- If you find a point that hinders your breathing, slowly move away; you can come back to it another time.
- The speed at which you decide to move determines the intensity of the exercise. The rule of thumb is: The slower you move, the more intense the feeling and, often, the more the tissue will relax.

8.5. TRAINING PLAN OR PLAN TO SUPPLEMENT EXISTING TRAINING

Please remember that in this chapter I am only talking about the myofascial RELEASE strategy and how to include it more effectively in a training context. In an actual training session I include other FReE strategies, which you can find in chapter 9. There are many ways in which myofascial release can be incorporated into your own program, some of which are listed below.

Remember that it is always a good idea to warm up with 2 to 5 minutes of aerobic exercise before a self-massage. Regular and targeted training regenerates myofascial tissue, promotes prolonged tension release, and leads to improved performance.

Like a massage, all the RELEASE exercises detailed here involve the application of pressure to the fascial tissue. This mechanically triggers an exchange of fluids in the fascia, which is literally squeezed like a sponge. These fluids also transport metabolic and lymph products with them. The fascia then fills with fluid again. This exchange stimulates the metabolism and improves the health of both the fascia and the surrounding organs. That is why it is important to revitalize and regenerate the fascia by applying pressure. This effect is achieved by osteopathic and physiotherapeutic techniques, as well as massages based on various techniques. Remember that tissue loves to be squeezed, requires well-measured pressure, and likes to slide.

Myofascial Release at the End of the Day

After an intense day full of stress that your posture often resents, the best thing to do is to take a bit of time for yourself. This will enable you to eliminate the tension that has built up during the day, restoring the correct length and balance of the muscles and fascia. I like to choose exercises that focus on the following areas, which are often susceptible to tension and muscle contractures. Choose your exercises carefully. What follows is an example of a full-body release sequence, starting from the feet.

- Foot
- Calf muscle
- Inner thigh
- Hip
- Pelvis
- Lumbar spine
- The entire spinal column
- Arm

Repetitions: 1 to 3 per exercise
Training session duration: 20 to 30 minutes, up to twice a week

I recommend that you drink plenty of water after this specific release training. Remember: Fascia and tissue need lots of fluids. This technique releases toxins. Metabolic waste is eliminated more easily with water.

RELEASE Exercises in a FReE Training Session

You will find myofascial RELEASE exercises in the following phases of a FReE training session.

- Warm-up
- Middle phase
- End phase

Repetitions: 1 to 3 per exercise
Training session duration: 30 to 55 seconds

RELEASE Exercises in a General Fitness or Sport-Specific Training Session
1. Warm-Up

- Start with light aerobic exercises to make sure that your body is sufficiently warm.
- Include exercises to eliminate tension.
- Choose exercises for all the myofascial lines, with varying rhythms, speeds, and angles. Start with slow movements before in-creasing the speed, but keep everything under control.

Repetitions: 1 to 3 per exercise
Warm-up duration: 5 to 20 minutes

This warm-up will also offer you other benefits if you take a few minutes to use it on a specific problematic area. In most athletes, the hip area (including the six internal rotators) and the lumbar spine are the most contracted regions and are key to all athletic movements. For total flexibility of the back, shoulders, and neck, the spinal column must be completely relaxed.

Before an intense strength training session, I prefer not to roll over a large area like the iliotibial band because it dehydrates the fascia and makes it unstable, and it needs time to recover and rehydrate.

2. Middle Phase
- Choose exercises to eliminate tension.
- Exercises can quickly loosen up a couple of trigger points during a training session.

3. End Phase
- Choose exercises to eliminate tension and relax.
- Help to optimize your recovery.

RELEASE Exercises in a Stretching/Pilates Session
- Include RELEASE exercises before a stretching session to make it more productive and relaxing.
- Include them in the warm-up and during the training session to eliminate tension.

- Choose relaxation exercises in the end phase.

Repetitions: 3 to 10

You will be surprised to see how much this improves your session.

8.6. CATEGORIZATION OF MYOFASCIAL RELEASE EXERCISES

In this chapter, the RELEASE exercises are grouped by myofascial lines: SBL – SFL – LL – DFL – AL.

8.6.1. SBL RELEASE

Rocking on the Sacrum With a Sensory Ball

Strategy
SBL RELEASE

Starting Position

Lie down on your back with your legs bent and your arms comfortably on the floor. Place a sensory ball in the middle of your sacrum.

Movement

Breathe in, move your pubis slightly toward your navel, and stretch and bend your lumbar spine toward the floor (retroversion of the pelvis). Breathe out and move your pubis away from your navel. Continue to rock back and forth, maintaining light pressure on the ball.

Focus On

- Fluid and effortless movements.
- Caution: Never place the ball under your coccyx.

Work On

- Release of tension in the sacrolumbar region.
- Stimulation of the parasympathetic autonomic nervous system.
- Stimulation of the tissue.
- Improving circulation.
- Increasing tissue oxygenation.

Discussion
Having performed the exercise, remove the ball and lower your pelvis onto the floor. Your sacral region should be relaxed, flat, and firmly on the floor.

Change Stimulus

Add the Balanced Breathing exercise on page 286 to increase the relaxation.

Knee Fall Out With a Sensory Ball

Strategy

Sacroiliac RELEASE

Starting Position

Lie down on your back with your legs bent and your arms comfortably on the floor. Place the ball in the middle of your sacrum and straighten your left leg, sliding your left buttock onto the floor. Move the ball slightly to the right, between the sacrum and buttocks (sacroiliac region).

Movement

Breathe in and move your right knee outward, away from the left iliac crest. Breathe out and allow your right leg to return to the starting position through elastic recoil. Repeat 10 times and then change sides.

Focus On

- Fluid and unforced movements.
- The buttock on the side of the bent leg must always stay on the floor.

Work On

- Relaxation of the sacroiliac region, SBL/LL.
- Opening of the femoral triangle (psoas, iliacus, and pectineus), DFL.
- Stimulation of the parasympathetic autonomic nervous system.

Discussion

Do you want to feel the effect of this exercise on the gluteus maximus? Having worked your right leg, remove the ball and lie on the floor. Can you feel it? Your right buttock should feel flat, like a flat tire. The feeling of relaxation is fantastic. This is just one side of the coin, but in this case both sides are positive.

At the front we have the opening of the femoral triangle (psoas, iliacus, pectineus, and neurovascular bundle, which pass under the inguinal ligament). These three muscles form a fan (which is why it is called the "femoral triangle"), which extends from the lesser trochanter to the hip bone and to the lumbar spine. Maintaining the tone and balanced length of these muscles gives freedom of movement and keeps the tissue healthy.

Who is it suitable for? This exercise is suitable for anyone with extreme hyperlordosis, who often has low back pain or tension, and for athletes in general. It will help you to loosen up, release tension, and relax.

8.6.2. SBL/LL RELEASE in the Sacrolumbar Region

Sideways Rocking on the Sacrum With a Sensory Ball

Strategy
Sacrolumbar RELEASE

Starting Position

Lie down on your back with your legs bent and your arms comfortably on the floor. Place one ball on the right side, between the sacrum and buttocks at middle pelvis height, and the other ball on the left side. With your legs together, lift them both off the floor and relax your feet.

Movement

Breathe in and move both your knees carefully to your left. Breathe out and return to the starting position. Repeat on the right. Continue to perform these movements. Once you have mastered the movement, carry on without stopping in the middle. Complete 5 to 8 movements per side.

Focus On

- Fluid and effortless movements.
- Caution: Never place the ball under your coccyx.
- Keep breathing (do not hold your breath).

Work On

- Release of tension in the sacrolumbar region.
- Stimulation of the parasympathetic autonomic nervous system.
- Stimulation of the tissue.
- Improving circulation.
- Increasing tissue oxygenation.

Discussion

If you notice a painful point as you move, stop, keep breathing, and try to loosen it on the ball. If the pain has subsided, continue with the exercise as described, gradually increasing the range of motion.

Change Stimulus

- Move the balls to the top of the iliac crest.
- Use a foam roller.

Back Massage Against the Wall With a Sensory Ball

Strategy
SBL RELEASE

Starting Position

Stand up straight with your back to the wall and wedge two balls between the wall and the right and left side of your spinal column, just below the shoulder blades.

Movement

Move your legs forward and start to roll slowly over the balls, lowering your pelvis toward the floor. Hold your back and pelvis in neutral position. Move slowly up and down as you continue to push against the balls, massaging your back. Repeat 5 to 8 times per region.

Focus On

- Pushing consistently against the balls.
- Maintaining your back and pelvis in neutral position.
- Breathing.

Work On

- Releasing the tension of the thoracic and lumbar spine.
- Improving flexibility.

Change Stimulus

- Use a foam roller.
- Lumbar spine massage: Place the balls in the lumbar region and start to roll from the bottom to the top.
- Loosen up on the sensory balls: This exercise improves breathing. Lie down on your back and place the two balls under the right and left side of your spinal column, just below the shoulder blades. Bring your arms above your head. If you feel too much tension or your back arches upward, move your arms down by your side. Hold the position, breathe, and let the balls loosen you up. If necessary, you could also move the balls to the lumbar region and repeat the exercise. Hold the position until you feel the tension diminish or completely abate.

Increase the Level of Difficulty

- Perform the exercise lying on your back. This will increase the pressure with your body weight. Begin with small movements between your shoulder blades and continue to breathe (without holding your breath). As you get more confident, gradually increase the range of motion until you get to your lower back but proceed delicately.

- Perform the same exercise with the foam roller. Place the foam roller under your shoulder blades and roll upward and then downward, using small movements. Keep breathing (do not hold your breath).

Strategy

Back of the neck SBL RELFASE

Starting Position

Lie on your back, with your legs bent. Place the foam roller under the back of your neck, as shown in the photo, and put your arms down by your side.

Movement

With delicate and slow movements, start to turn your gaze and your head to the right. Use these small movements to look for trigger points. Once you have found one, stop and apply gentle pressure or delicately massage. Switch direction.

Focus On

Slow and delicate movements.

Work On

- Releasing tension in the back of the neck (this exercise is for people who often have a sore neck or a headache).
- Stimulation of the parasympathetic autonomic nervous system.

Discussion

In terms of frequency, you can roll 5 to 8 times every day. This exercise is most beneficial if done in the evening lying on your back, as it stimulates the autonomic system, activates the parasympathetic system, and makes you sleepy.

Reduce the Level of Difficulty

Stand up and perform the exercise with the foam roller against the wall.

8.6.3. SFL RELEASE

Leg Lift in Prone Position With a Sensory Ball

Starting Position

Lie down on your front with the ball under the left side of your pelvis (between the crest and the anterior superior iliac spine). Rest your forehead comfortably on the back of your hands, activating the kite.

Movement

Straighten and then slowly lift your left leg into the air, keeping your pelvis resting firmly on the ball and the floor. Return to the starting position. Repeat 3 times.

With your leg still raised, bend your knee to bring your heel toward your buttocks and then straighten your leg again without resting it on the floor. Repeat 3 times.

Move your straight leg away from the midline of your body and then return to the starting position without resting it on the floor. Repeat 3 times.

With your leg out to the side, bend your knee to bring your heel toward your buttocks and then straighten your leg again without lowering it to the floor. Repeat 3 times.

Return to the starting position, bend your knee to bring your heel toward your buttocks, straighten your leg, and rest it on the floor. Repeat the entire exercise with your right leg.

Focus On

- Always maintaining contact with the ball and the floor.
- Active kite.
- Spine in a neutral position.

Work On

- Releasing tension from the front of your pelvis on the SFL/LL.
- Improving the flexibility of the rectus femoris and tensor fasciae latae muscles.
- Strengthening the hamstrings and gluteal muscles.

Change Stimulus

When you are in the position shown below, simply move your leg away from your midline and then back in toward the midline, increasing the work of the internal and external rotators of the hip.

Strategy
SFL RELEASE

Starting Position

Lie down on your front, with your feet together and the ball under the middle of your sternum. Rest your arms on the floor, with your elbows at shoulder height. Lengthen your neck to activate the kite.

Movement

Breathe in and push the ball forward slightly with your sternum, extending and raising your chest and head in the same direction. Hold the position and then return to the starting position the next time you exhale. Repeat 5 to 8 times.

Focus On

- Active kite.
- Looking for the stretch.

Work On

- Releasing tension from the sternum, SFL/ Superficial Arm Line.
- Improving the mobility and flexibility of the chest and shoulders.

Discussion

Once you have completed the exercise, remove the ball and lie down on the floor. Your chest area will be much more relaxed.

Sliding Over the Sternum With a Sensory Ball in a Seated Position

Strategy
SFL RELEASE

Starting Position

Sit down comfortably on the floor or on a chair and pick up the ball with your right hand. Activate the kite and place the ball against the lower left side of your sternum.

Push the ball gently against your skin (so that the ball "grabs" the skin) and slowly slide the ball all the way up to your clavicle, pushing gently against your skin as you go. Lift the ball off your skin and return to the lower left side of your sternum. Repeat 3 times and then move to the lower right side.

Focus On
- Active kite.
- Sliding the ball slowly.

Work On
- Releasing tension from the sternum, SFL/Superficial Arm Line.
- Improving the mobility and flexibility of the chest and shoulders.

Discussion

Once you have completed the exercise on the left side, put the ball down and "listen". Do you notice a difference between your right and left sides? The left side may feel longer, and, if you look in the mirror, your shoulders may not be the same height. The good thing about this exercise is that it gives you a positive postural input.

The feeling of relaxation that comes from sliding the ball cannot be reproduced by simply pressing the ball against the sternalis, or by rolling. Sliding leaves a trace in the structures: Show them the direction in which they should go.

Take a look around at how many people have hyperkyphosis. For many people with a kyphotic posture, their shoulders are closed forward and their sternum moves toward their pelvis. These cases are ideal candidates for this sliding exercise. As you apply pressure, the ball "grabs" the skin. As you slowly push the ball up, you should feel it pulling the skin and encountering stiffer points that prevent it from continuing. Stop at these points until you feel the tissue relaxing, enabling you to continue sliding the ball upward. This is an effective way to stretch the fascia. Remember: Fascia adapts to any demand.

Test

After you slide the ball on your left side and before you do the right side, move your arms in circles. You should normally notice a striking difference between the two sides. The side that has been massaged should feel lighter, with greater shoulder mobility.

8.6.4. LL RELEASE

Circles Around the Piriformis With the Duo Ball

Starting Position

Lie down on your back and place the duo ball under your right side, between the femoral head and the ilium. Straighten your left leg on the floor and bend and raise your right leg. Spread your arms comfortably on the floor.

Breathe in and begin the circling motion by moving your right knee outward.

At maximum abduction of the femur, turn your leg inward and rest the inside of your foot on the floor.

Breathe out and continue the circle with the inside of your foot on the floor.

Complete the circle by returning to the starting position.

Repeat the circle 5 times and then repeat the whole exercise on the other side.

Focus On

Fixed point: pelvis; mobile point: knee.

Work On

- Releasing the tension of the piriformis.
- Increasing mobility and flexibility.

Change Stimulus

Increase the size of the circle: Raise your knee as close to your chest as you can, then outward toward your armpit, then turn your leg inward and complete the circle, without moving your pelvis since the goal is to mobilize your hip.

Discussion

This is a clever relaxation exercise of the piriformis. What do I mean by this? I often see the ball placed in the middle of the buttocks on the piriformis muscle, which is an important pathway of the sciatic nerve. This will compress or constrict the sciatic nerve. However, placing the duo ball as shown in this exercise will relax the part of the muscle and the tendon close to the femoral head, which will stimulate without constricting the sciatic nerve. Clever, isn't it?

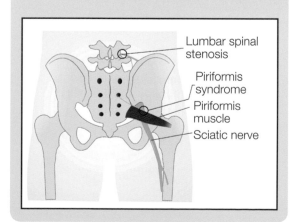

Lumbar spinal stenosis

Piriformis syndrome

Piriformis muscle

Sciatic nerve

Sitting on a Sensory Ball in the "Z" Position

Strategy
Piriformis muscle RELEASE

Starting Position

Sit in the "Z" position, with your right leg in front and your left leg turned to the side. Place the ball between the femoral head and ilium (laterally between your thigh and pelvis) and straighten your spine with your arms straight and down by your side.

Movement

Straighten and then tilt your upper body to the right and raise your left arm to increase the pressure on the ball. To come back up, start the movement from your ribs (they should be the first things to move), with your arm and head following. Repeat 5 to 8 times and then change sides.

Focus On

Fixed point: pelvis; mobile point: arm.

Work On

*   Releasing the tension of the piriformis.
*   Increasing hip mobility and flexibility.

Discussion

This exercise may make it easier to sit in the "Z" position (see exercises like the one on page 193).

Strategy
LL RELEASE – iliotibial tract

Starting Position

Lie on your side with your forearm on the floor and your shoulder perpendicular to your elbow. Place the ball under your right side, with your right leg straight and resting on the floor. Bend your left leg, with your foot on the floor in front of your right leg.

Movement

Start to move the ball slightly forward toward your pubis and then back toward your buttocks. As the ball moves back and forth, it should also start moving slowly toward your knee. Continue this zig-zag motion until the ball almost reaches your knee.

At this point turn your body so that you are face down, moving the ball slightly to the outside of your right thigh (exactly between the vastus lateralis of the quadriceps and the iliotibial tract). Slowly roll on the ball toward your pelvis until it reaches the anterior superior iliac spine. Complete the whole exercise 3 times and then repeat on the other side.

Focus On

- Fluid and effortless movements.
- Holding the kite position.
- Breathing (do not hold your breath).

Work On

- Releasing the tension on the LL – iliotibial tract.
- Improving sliding between the iliotibial tract and the quadriceps/hamstrings.
- Increasing flexibility of the hamstrings (back of the thigh).

- Improving circulation.
- Increasing tissue oxygenation.
- Increasing strength.

Discussion

It is certainly true that it is not the most comfortable position. My advice is to maintain an active kite; otherwise you will load the back of your neck.

How does this exercise increase your strength? Imagine different layers of muscle and fascial tissue glued and pressed together. Do you think their movements would be fluid? Is it possible to strengthen these muscles? It certainly is but only to a limited extent. As we go up we are trying to separate the iliotibial tract from the lateral quadriceps, while the zig-zagging reduces the tension of the iliotibial tract. By performing this exercise, the work between various muscles and layers of tissue (quadriceps, iliotibial tract, hamstrings) could significantly improve in terms of connections, strength, and sliding between the muscles.

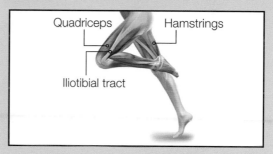

Who is it suitable for? It is not suitable for people with varicose veins.

Change Stimulus

Tension often builds up on the LL around the head of the fibula, which in turn leads to knee tension. If you place the ball just below the outside of the knee, you will find a protruding bone (exactly at the head of the fibula). Press gently for 3 seconds and release; repeat 5 to 8 times.

Reduce the Level of Difficulty

If this exercise doesn't suit you, you can do it standing up sideways against the wall. Place the ball in a sock and hold the end with your right hand. Pass the stocking over your right shoulder. Wedge the ball between the side of your pelvis and the wall. Push your body weight against the ball and start to move it slowly forward and backward, massaging the LL. The more you move the ball down, the more you have to stretch the stocking. This method will stop you from losing the ball.

8.6.5. DFL RELEASE

Circles With the Sensory Ball on Your Cheek

Strategy
DFL masseter muscle RELEASE

Starting Position

Sit down comfortably on the floor or on a chair, hold the ball in your right hand, activate the kite, and hold the ball against your right cheek.

Movement

Press gently and move the ball in circular motions 5 to 10 times to massage your cheek and jaw then switch sides.

Focus On

- Active kite.
- Breathing.

Work On

- Releasing the tension of your cheek and neck.
- Releasing the tension of the masseter muscle, DFL.

Discussion

To reduce tension caused by stress, we sometimes clench our teeth, and this can happen both during the day and at night. This fatigues the masticatory muscles, which causes contractures and pain. This is one of many exercises that may help to release built-up tension.

Who is it suitable for? This exercise is suitable for anyone who clenches his or her teeth and anyone with neck pain.

8.6.6. AL RELEASE

Shoulder Movements Against the Wall With a Sensory Ball

Strategy
RELEASE of the shoulders and the Superficial Back Arm Line

Starting Position

Stand sideways against the wall with your legs hip-width apart. Wedge the ball between your shoulder and the wall. Push gently against the ball. Hold the kite position.

Movement

Slowly move the ball (move it half an inch [1 cm], hold for one breath, and continue for another half an inch, etc., first forward and then backward, massaging the whole shoulder region. Slow down and move the ball less than half an inch each time if you find a painful point. When the tension diminishes, continue the movement 5 to 8 times per region.

Focus On

- Active kite.
- Slow movements.

Work On

- Releasing tension in the shoulder region.
- Improving mobility, flexibility, and strength.

Discussion
To keep the tissue flexible, repeat this exercise 2 or 3 times a week.

Change Stimulus

Move the ball upward and then downward.

Shoulder Blade Movements Against the Wall With a Mini Foam Roller

Strategy
RELEASE of the lateral edge of the shoulder blade

Starting Position
Stand sideways against the wall with your legs hip-width apart. Raise your right arm above your head and wedge the mini foam roller between the lateral edge of your shoulder blade and the wall. Push gently against the mini foam roller.

Movement
Try to roll up and down slowly on the most sensitive or painful points. If you find a painful point, make very small movements in this region. If you cannot find any trigger points, try moving the ball over your shoulder blade. This tends to be a region where tension builds up. The pain should diminish after 5 to 8 repetitions and disappear completely after 4 to 5 training sessions.

Focus On
Pushing constantly against the mini foam roller.

Work On
Eliminating tension in the lateral region of the shoulder blade.

Discussion
The myofascial structures found above and below the shoulder blade are stressed and undergo changes when sitting for prolonged periods. That is why tension predominantly builds up in this region.

Change Stimulus
- The further your legs are from the wall, the more your body weight on the mini foam roller will increase and the harder the exercise will be. You can use this technique to adjust the intensity.
- Use a large foam roller.
- Increase the intensity: Perform the whole exercise lying on your side with the outer edge of your shoulder blade on the foam roller. Perform small movements.

Forearm Movements Against the Wall With a Mini Foam Roller

Strategy
Forearm RELEASE

Starting Position
Stand close to the wall with your legs hip-width apart. Wedge a mini foam roller between the wall and your forearm, with your elbow at shoulder height.

Movement
Slowly roll your forearm upward (toward your hand) and back down again. Look for painful points and perform small movements on any painful regions you find. In these positions, open and close your hand or flex and extend your wrist.

Focus On
- Active kite.
- Pushing constantly against the mini foam roller.

Work On
Relaxing the finger flexors.

Discussion
Anyone who spends hours in front of the computer can do this exercise every day or at least twice a week. It is very effective at preventing or treating so-called tennis elbow and for anyone who works long hours at the computer and uses the mouse a lot.

Forearm Movements on All Fours With a Foam Roller

Strategy
Arm Lines RELEASE

Starting Position

Get on all fours, place your right hand on the foam roller, and activate the kite.

Movement

Breathe in and stretch forward. Press down on the foam roller as you extend your fingers upward. Breathe out and return to the starting position. Repeat 5 to 8 times.

Focus On

- Active kite.
- Pressing down constantly on the foam roller.

Work On

Relaxing the finger flexors.

282

Strategy
Arm Lines RELEASE

Starting Position

Get on all fours, place your right arm on the foam roller with the palm of your hand facing up, and activate the kite.

Movement

Turn your upper body and slowly roll your arm toward your shoulder and back toward your hand. Look for painful points and perform small movements on any painful regions you find. Repeat 5 to 8 times and then change arms.

Focus On

- Active kite.
- Pushing constantly against the foam roller.

Work On

Releasing the lines that affect the arm and shoulder.

Pectoral Movements Lying on Your Front

Starting Position

Lie down on your front, place the ball under your pectoral muscle, and extend your arms outward at shoulder height.

Movement

Slowly move on the ball toward your shoulder in search of trigger points. When you find one, perform small movements and try to reduce the tension.

Focus On

Breathing.

Work On

Releasing the Front Arm Lines.

Reduce the Level of Difficulty

Pectoral movements against the wall: Perform the whole exercise against the wall.

8.6.7. Breathing RELEASE

5 to 10 Balanced Breathing With Weight

> **Strategy**
> Stimulate the parasympathetic system

Starting Position and Movement

Lie down on your back, with your legs bent, and rest your arms on the floor by your sides. Place a 2- to 11-pound (1 to 5 kg) weight on your abdominal wall (this is not a strength exercise).

Movement

Breathe in through your nose for 5 seconds and out through your mouth for 10 seconds. Repeat a further 10 to 15 times.

Focus On

- Breathing in toward the lumbar spine.
- The weight should hardly move as you breathe in.

> **Discussion**
> You can also perform this exercise without the weight. The weight helps you to relax.

Work On

Stimulating the parasympathetic nervous system.

Balanced Breathing

Strategy

Stimulate the parasympathetic system

Starting Position and Movement

Lie down on your back and put your hands on your abdomen or down by your sides. Breathe in through your nose for 5 seconds, hold your breath for 1 second, breathe out through your mouth for 5 seconds, and hold for 1 second. Repeat for 10 to 15 breathing cycles (one cycle consists of breathing in and breathing out).

Focus On

Breathing in toward the lumbar spine.

Work On

- Synchronizing the heart and breathing.
- Balancing heart rate variability.

Discussion

This exercise synchronizes the heart and breathing and stimulates the hormone DHEA (a cortisol antagonist), and the cortisol in the blood is eliminated. You will feel more relaxed, and your mood will improve. As we age, our DHEA and cortisol curves tend to flatten out, which means that our body finds it difficult to produce these two hormones. This has a negative impact on the rhythm of other hormone systems and on the immune system. Techniques like yoga and meditation help to harmonize the relationship between these two hormones.

TRAINING PROGRAMS

9

After so much theory and education, we finally come to the training itself.

In this chapter I propose a range of training program examples with different focuses; all you have to do is choose your goal. As we know, the fascial structures are all different, and so too are my training proposals.

Test and verify any changes or improvements; all feedback is welcome.

Step one: Choose your area of focus.

- Healthy feet (heel spur/hallux valgus)
- Carpal tunnel
- Strong shoulders
- FReE neck
- Flabby thighs and cellulite
- For running
- To improve flexibility
- To improve mobility
- Integration into Pilates
- Integration into functional training
- FReE training
- Fascia-nating moves FReE Beast

Step two: Choose the training program that is most suited to your training regime. Follow it diligently, and you will be surprised at how quickly you progress.

Step three: Never work with pain.

Step four: Perform the listed exercises in each program according to the Training and Homework guidance listed in the Training Guidelines section after the programs.

The programs also include functional training and Pilates exercises that are not covered in this book. These exercises are written in lower case.

Note: You can also include the exercises from the various programs in other types of training to fully meet your training needs.

9.1. PROGRAMS FOR HEALTHY FEET: HEEL SPUR AND HALLUX VALGUS

Concept: Non-stop program and sequence of exercises.
Structure: One sequence of 6 to 15 exercises constitutes a training program.
Focus On: Balancing tension in the lines of the lower limbs (SBL-SFL-LL-DFL). Eliminating pain, improving functionality and athletic performance.
Strategies: RELEASE, FEEL IT, STRETCH, ENERGY, and MOBILITY.

HEEL SPUR (CALCANEAL SPUR) – PROGRAM 1	HEEL SPUR (CALCANEAL SPUR) – PROGRAM 2
EQUIPMENT: **SENSORY BALL - FOAM ROLLER - PLYO BOX**	**EQUIPMENT: SENSORY BALL**

HEEL SPUR (CALCANEAL SPUR) – PROGRAM 1

EQUIPMENT:
SENSORY BALL - FOAM ROLLER - PLYO BOX

FOOT MASSAGE WITH BALL
1 x 30 sec per foot (page 60)

ANTERIOR STIMULATION OF
THE ANKLES RESTING ON THE
FOAM ROLLER 1 x 5 (page 95)

POSTERIOR STIMULATION OF
THE ANKLES RESTING ON THE
FOAM ROLLER 1 x 5 (page 97)

PRESSURE AND MASSAGE ON
THE CALF MUSCLE WITH A
FOAM ROLLER 1 x 5 per leg
(page 98)

PRESSURE AND MASSAGE ON
THE INNER LEG WITH A FOAM
ROLLER 1 x 5 per leg (page 99)

PRESSURE AND MASSAGE
ON THE INNER THIGH WITH A
FOAM ROLLER
1 x 5 per leg (page 100)

ON/OFF SHORTENED HIP
FLEXOR 1 x 5 per side (page 90)

ON/OFF HIP FLEXOR STRETCH
3 x 3 sec per leg (page 91)

ELASTIC BOUNCES 3 x 10 sec
(page 237)

ELEPHANT ON PLYO BOX
1 x 5–10 per side (page 211)

MODIFIED WARRIOR ON PLYO
BOX 1 x 5–10 per side
(page 213)

LEG LIFT IN PRONE POSITION
WITH SENSORY BALL
1 x series per side (page 272)

ELASTIC BOUNCES 3 x 10 sec
(page 237)

HEEL SPUR (CALCANEAL SPUR) – PROGRAM 2

EQUIPMENT: SENSORY BALL

FOOT MASSAGE WITH BALL
1 x 30 sec per foot (page 60)

ELASTIC BOUNCES 3 x 10 sec
(page 237)

STEPPING BACK AND FORTH
1 x 10 per leg (page 109)

"L" STEP 1 x 10 per leg
(page 110)

ROUND ELEPHANT
1 x 10 per leg (page 156)

ZIG ZAG ON THE LATERAL LINE
WITH A SENSORY BALL
1 x 3 per side (page 277)

THREE FEET IN MOVEMENT
1 x 5 per side (page 163)

SUPPORTED LUNGE WITH BOW
1 x 5 per side (page 235)

BOOMERANG IN SIDEWAYS
HALF-KNEELING POSITION
1 x 5 per side (page 188)

BEACHED STORK – KNEE RAISE
1 x 5 per side (page 123)

WINDMILL IN HALF-KNEELING
POSITION 1 x 5 per side
(page 184)

LIZARD 1 x 20 sec per leg
(page 189)

LARGE CORKSCREW
3 x 15 sec per leg (page 197)

SIDEWAYS BOX
3 x 15 sec per leg (page 217)

Training Guidelines

Perform the exercises barefoot and distribute your body weight over the three contact points of the foot.

Training: 2 or 3 times a week for 2 weeks and then move on to program 2.

Homework: Find some time every evening to massage your feet with the ball.

Note: You can also include the exercises from the various programs in other types of training to fully meet your training needs.

HALLUX VALGUS – PROGRAM 1

EQUIPMENT: SENSORY BALL
Goal: To prevent and improve hallux valgus at an early stage

FOOT MASSAGE WITH BALL
1 x 20 sec per foot (page 60)

STEPPING BACK AND FORTH
1 x 10 per leg (page 109)

"L" STEP 1 x 10 per leg (page 110)

FRONTAL WOODPECKER: MOVEMENTS IN HALF-KNEELING POSITION 1 x 5 per leg (page 114)

STORK AGAINST THE WALL
5 x 3 sec per leg (page 82)

SHIFT SQUAT 1 x 5 per side
(page 181)

MOVING STORK 1 x 5–10 per leg
(page 119)

THREE FEET IN MOTION 5 x leg
(page 161)

THORACIC SPINE ROTATIONS ON TWO POINTS OF SUPPORT
1 x 5 per side (page 129)

Note: You can also include the exercises from the various programs in other types of training to fully meet your training needs.

HALLUX VALGUS – PROGRAM 2

EQUIPMENT: SENSORY BALL – FOAM ROLLER – RESISTANCE BAND
Goal: To prevent and improve hallux valgus at an early stage

FOOT MASSAGE WITH BALL
1 x 20 sec per foot (page 60)

ANTERIOR STIMULATION OF THE ANKLES RESTING ON THE FOAM ROLLER 1 x 10 (page 95)

POSTERIOR STIMULATION OF THE ANKLES RESTING ON THE FOAM ROLLER 1 x 10 (page 97)

PRESSURE AND MASSAGE ON THE CALF MUSCLE WITH A FOAM ROLLER 1 x 5 per leg (page 98)

PRESSURE AND MASSAGE ON THE INNER LEG WITH A FOAM ROLLER 1 x 5 per leg (page 99)

STEPPING BACK AND FORTH WITH A RESISTANCE BAND
1 x 10 per side (page 110)

SIDE STEP AND DIAGONAL STEP
1 x 10 per side (page 111)

CRAB IN SQUAT POSITION
1 x 3–5 per side (page 117)

RUNNER'S KNEE IN SQUATTING POSITION 1 x 3–5 per side
(page 118)

SHIFT SQUAT IN HALF-KNEELING POSITION 1 x 5–8 per side
(page 182)

UNCOMFORTABLE HALF-SITTING POSITION 3 x 5 per leg
(page 169)

THREE-LEGGED WINDMILL
1 x 5 per side (page 186)

KNEE SAVER-LATERAL
STABILIZATION
1 x 10 per side (page 120)

SUPINE CORKSCREW
1 x 5 per side (page 196)

Note: You can also include the exercises from
the various programs in other types of training
to fully meet your training needs.

HALLUX VALGUS – PROGRAM 3
EQUIPMENT: RESISTANCE BAND **Goal:** To prevent and improve hallux valgus at an early stage

ELASTIC BOUNCES 3 x 10 sec
(page 237)

STEPPING BACK AND FORTH
WITH A RESISTANCE BAND
1 x 10 per side (page 110)

SIDEWAYS LUNGE 1 x 5 per leg
(page 112)

DIAGONAL LUNGE 1 x 5 per leg
(page 112)

UNCOMFORTABLE HALF-SITTING
POSITION 3 x 5 per leg (page 169)

ELEPHANT IN MOTION
1 x 5 per leg (page 157)

BEETLE STRETCH 1 x 5–10
(page 218)

SPREADING YOUR KNEES IN
HALF-SITTING POSITION
1 x 3–5 per side (page 115)

GORILLA 1 x 5 (page 214)

STATUE 1 x 20 sec per leg
(page 174)

SHIFT SPLIT SQUAT
1 x 5–7 per leg (page 182)

PRONE CORKSCREW
1 x 5 per side (page 199)

SPREADING YOUR TOES
1 x 10–20 sec (page 166)

Training Guidelines
Perform the exercises barefoot and distribute your
body weight over the three contact points of the
foot.

Training: 2 or 3 times a week for 2 weeks and then
move on to the next program.

Homework: Every time your phone rings or you
receive a text, check that your body weight is
evenly distributed over the three contact points
of your feet.

Note: You can also include the exercises from
the various programs in other types of training
to fully meet your training needs.

9.2. PROGRAMS FOR CARPAL TUNNEL AND THORACIC OUTLET SYNDROME (TOS)

Concept: Non-stop program and sequence of
exercises.
Structure: One sequence of 6 to 15 exercises
constitutes a training program.
Focus On: Balancing tension in the Arm Lines.
Eliminating pain, improving functionality and
athletic performance.
Strategies: RELEASE, FEEL IT, MOBILITY, and
STRETCH.
Discussion: This method can be used by
anyone to improve

- posture,
- carpal tunnel and thoracic outlet syndrome,
- the position of the shoulders (internally
  rotated),
- hand support on all fours, and
- the following workout.

Use these programs to warm up before
strength training, kettlebell training, functional

training, etc. to improve performance and prevent injury.

The first two programs are designed to relieve tension associated with inflammation and to balance the Arm Lines. Programs 3 and 4 are designed to balance the Arm Lines by strengthening and stretching the weak links.

CARPAL TUNNEL and TOS – PROGRAM 1
EQUIPMENT: SENSORY BALL – FOAM ROLLER

EYES ALERT, NECK LOOSE (page 87)

BACK MASSAGE AGAINST THE WALL WITH A SENSORY BALL (page 270)

FOREARM MOVEMENTS AGAINST THE WALL WITH A MINI FOAM ROLLER (page 281)

SHOULDER MOVEMENTS AGAINST THE WALL WITH A SENSORY BALL (page 279)

PECTORAL MOVEMENTS AGAINST THE WALL WITH A SENSORY BALL (page 284)

CIRCLES WITH THE SENSORY BALL ON YOUR CHEEK (page 278)

CHEST LIFT IN PRONE POSITION WITH A SENSORY BALL (page 273)

THORACIC SPINE ROTATIONS (page 130)

LOOSEN UP ON THE SENSORY BALLS (page 270)

KNEE FALL OUT WITH A SENSORY BALL (page 268)

Training Guidelines

Repeat each exercise 5–10 times.

Keep breathing (do not hold your breath).

Training: Perform the training session every other day for a total of four training sessions and then move on to the next program.

Homework: Eyes Alert, Neck Loose (page 87), every day.

Note: You can also include the exercises from the various programs in other types of training to fully meet your training needs.

CARPAL TUNNEL and TOS – PROGRAM 2
EQUIPMENT: SENSORY BALL – FOAM ROLLER

SLIDING OVER THE STERNUM WITH A SENSORY BALL IN A SEATED POSITION (page 274)

LOOSEN UP ON THE SENSORY BALLS (page 270)

PECTORAL MOVEMENTS LYING ON YOUR FRONT (page 284)

BACK MASSAGE WITH A FOAM ROLLER (page 270)

FOREARM MOVEMENTS ON ALL FOURS WITH A FOAM ROLLER (page 282)

ARM MOVEMENTS ON ALL FOURS WITH A FOAM ROLLER (page 283)

SPLITS AGAINST THE WALL 1 x 30–45 sec (page 220)

Training Guidelines

Repeat each exercise 5–10 times.

Keep breathing (do not hold your breath).

Training: Perform the training session every other day for a total of four training sessions and then move on to the next program.

Homework: Eyes Alert, Neck Loose (page 87) and Circles With Your Wrists, every day.

Note: You can also include the exercises from the various programs in other types of training to fully meet your training needs.

CARPAL TUNNEL and TOS – PROGRAM 3	CARPAL TUNNEL and TOS – PROGRAM 4
EQUIPMENT: RESISTANCE BAND – CLUBS – PLYO BOX	**EQUIPMENT: WEIGHTS**

ALTERNATING SWING WITH WOODEN CLUBS (page 258)

ARMS TO THE WALL 1 x 30 sec (page 79)

EYES ALERT, NECK LOOSE (page 87)

SPIDER WOMAN (page 131)

THORACIC SPINE ROTATIONS (page 130)

GIRAFFE: SWINGING WITH RESISTANCE BAND (page 225)

SIDEWAYS SPIDER AGAINST THE WALL (page 222)

CIRCLES WITH YOUR ARM AGAINST THE WALL (page 138)

AIRCRAFT MARSHALLING WITH CLUBS (page 142)

SIDEWAYS TILTING OF THE UPPER BODY IN STANDING POSITION (page 126)

BIRD DOG AGAINST THE WALL 2 x 30 sec per arm (page 176)

SIDEWAYS MODIFIED WARRIOR ON PLYO BOX 3 x 3 sec (page 224)

Training Guidelines
Repeat each exercise 5–10 times.

Keep breathing (do not hold your breath).

Training: Perform the training session every other day for a total of four training sessions and then move on to the next program.

Homework: Spider Woman (page 85), every day

Note: You can also include the exercises from the various programs in other types of training to fully meet your training needs.

ALTERNATING SWING WITH SOFT KETTLEBELLS (page 258)

DYNAMIC LITTLE MERMAID (page 193)

PRONE LARGE CORKSCREW (page 201)

SPIDER AGAINST THE WALL 3 x 15 sec (page 221)

GIRAFFE: SWINGING WITH WEIGHTS (page 225)

BAMBOO AGAINST THE WALL (page 205)

BIRD DOG AGAINST THE WALL (page 176)

WINDMILL AGAINST THE WALL (page 208)

CRAB (page 168)

SUPINE CORKSCREW (page 196)

Training Guidelines
Repeat each exercise 5–10 times.

Keep breathing (do not hold your breath).

Training: Perform the training session every other day for a total of four training sessions.

Homework: Spider Against the Wall (page 221), every day

Note: You can also include the exercises from the various programs in other types of training to fully meet your training needs.

9.3. PROGRAMS FOR STRONG SHOULDERS

Concept: Non-stop program and sequence of exercises.

Structure: One sequence of 6 to 15 exercises constitutes a training program.

Focus On: Balancing the Arm Lines. Improve posture, everyday functionality, and athletic performance.

Strategies: FEEL IT, RELEASE, STRETCH, ENERGY, and MOBILITY.

SHOULDERS – PROGRAM 1
EQUIPMENT: SENSORY BALL – FOAM ROLLER **Goal:** FEEL IT – feel the Arm Lines

MOVING STORK 1 x 5 per side (page 119)

BACK MASSAGE AGAINST THE WALL WITH A SENSORY BALL 1 x 10 (page 270)

BACK TO THE WALL IN STANDING POSITION 3 x 20 sec (page 81)

SHOULDER MOVEMENTS AGAINST THE WALL WITH A SENSORY BALL 1 x 10 per side (page 279)

FOREARM MOVEMENTS AGAINST THE WALL WITH A MINI FOAM ROLLER 1 x 10 per side (page 281)

PECTORAL MOVEMENTS AGAINST THE WALL WITH A SENSORY BALL 1 x 10 per side (page 284)

SHOULDER BLADE MOVEMENTS AGAINST THE WALL WITH A MINI FOAM ROLLER 1 x 10 per side (page 280)

SPIDER WOMAN 1 x 5–7 per side (page 131)

CIRCLES WITH YOUR ARM AGAINST THE WALL 1 x 10 per arm (page 138)

THE SHOULDER BLADES WRAP 1 x 5–8 (page 133)

WINDMILL AGAINST THE WALL 1 x 5 per side (page 208)

SPIDER AGAINST THE WALL 3 x 15 sec (page 221)

KNEE SAVER-LATERAL STABILIZATION 1 x 10 per side (page 120)

CRAB 5 x 15 sec (page 168)

Training Guidelines

Perform the exercises correctly and in a controlled manner (active kite).

Training: Perform the training session every other day for a total of four training sessions and then move on to the next program.

Note: You can also include the exercises from the various programs in other types of training to fully meet your training needs.

SHOULDERS – PROGRAM 2
EQUIPMENT: RESISTANCE BAND – FOAM ROLLER – WEIGHTS – CLUBS **Goal:** To balance the Arm Lines

EYES ALERT, NECK LOOSE 1 x 10 per side (page 87)

CIRCLES WITH THE RESISTANCE BAND 1 x 10 per side (page 141)

FOREARM MOVEMENTS ON ALL FOURS WITH A FOAM ROLLER 1 x 10 per side (page 282)

ARM MOVEMENTS ON ALL FOURS WITH A FOAM ROLLER 1 x 10 per side (page 283)

PECTORAL MOVEMENTS LYING ON YOUR FRONT 1 x 5 per side (page 284)

SHOULDER BLADE MOVEMENTS ON THE FLOOR WITH A FOAM ROLLER 1 x 5 per side (page 280)

ARMS TO THE WALL 1 x 5 sec per position (page 79)

INFINITE SWING WITH KETTLEBELL 1 x 10 per arm (page 260)

GIRAFFE: SWINGING WITH WEIGHTS 1 x 5–10 (page 225)

SIDEWAYS TILTING OF THE UPPER BODY IN STANDING POSITION 1 x 10 per side (page 126)

AIRCRAFT MARSHALLING WITH CLUBS 1 x 5–10 (page 142)

BIRD DOG AGAINST THE WALL 1 x 30 sec per leg (page 176)

DANCING SPIDER AGAINST THE WALL (page 223)

SIDEWAYS SPIDER AGAINST THE WALL 5 x 3 sec (page 222)

CRAB 4 x 10 sec (page 168)

Training Guidelines

Perform the exercises correctly and in a controlled manner (active kite).

Training: Perform the training session every other day for a total of four training sessions and then move on to the next program.

Note: You can also include the exercises from the various programs in other types of training to fully meet your training needs.

SHOULDERS – PROGRAM 3
Goal: To strengthen the Arm Lines

Repeat program 2 and add the following exercises:

TRANSVERSE SEATED WINDSHIELD WIPERS 1 x 5 per side (page 125)

GECKO PUSH-UP AGAINST THE WALL 1 x 5–10 (page 250)

CRAB 2 x 3 sec for each hand position (page 168)

CRICKET PUSH-UP (page 247)

Training Guidelines

Perform the exercises correctly and in a controlled manner (active kite).

Training: Perform the training session every other day for a total of four training sessions.

Note: You can also include the exercises from the various programs in other types of training to fully meet your training needs.

9.4. PROGRAM FOR THE NECK

Concept: Non-stop program and sequence of exercises.

Structure: One sequence of 6 to 15 exercises constitutes a training program.

Focus On: Alleviating neck tension.

Strategies: FEEL IT, RELEASE, STRETCH, ENERGY, and MOBILITY.

FReE NECK
EQUIPMENT: SENSORY BALL – FOAM ROLLER – PLYO BOX **Goal**: To alleviate neck tension

FOOT MASSAGE WITH BALL 1 x 30 sec per foot (page 60)

BACK MASSAGE WITH A FOAM ROLLER 1 x 10 (page 270)

BACK TO THE WALL IN STANDING POSITION 3 x 20 sec (page 81)

EYES ALERT, NECK LOOSE 1 x 5 (page 87)

SPIDER WOMAN 1 x 5 per side (page 131)

PRESSURE AND MASSAGE ON THE CALF MUSCLE WITH A FOAM ROLLER 1 x 5 per leg (page 98)

SLIDING OVER THE STERNUM WITH A SENSORY BALL IN A SEATED POSITION 3 per side (page 274)

PECTORAL MOVEMENTS AGAINST THE WALL WITH A SENSORY BALL 3 x 4 per side (page 284)

CIRCLES WITH YOUR ARM AGAINST THE WALL 1 x 10 per arm (page 138)

BAMBOO AGAINST THE WALL 3 x 5 sec per side (page 205)

ELEPHANT ON THE PLYO BOX 1 x 5 per side (page 211)

DYNAMIC CAMEL 1 x 5 with platform under your hands (page 160)

ROCKING ON THE SACRUM WITH A SENSORY BALL 1 x 5–10 (page 267)

BACK OF THE NECK RELEASE WITH A FOAM ROLLER 1 x 3–5 per side (delicately) (page 271)

Training Guidelines

Perform the exercises correctly and in a controlled manner.

Training: Perform the training session every other day for a total of four training sessions and then move on to the next program.

Note: You can also include the exercises from the various programs in other types of training to fully meet your training needs.

9.5. PROGRAMS FOR FLABBY THIGHS AND CELLULITE

What Are Flabby Thighs?

They are caused by a form of cellulite that builds up between the buttocks and the outer thigh, unbalancing the body. Combatting this disproportionate build-up of fat is not easy but also not impossible. To achieve a satisfying outcome, you have to work on multiple fronts:

- Correct diet
- Correct posture
- Targeted exercises

What Can Cause Flabby Thighs?

- Being overweight
- A poor diet, excessive alcohol consumption
- Sedentary lifestyle
- Incorrect posture
- Low fluid intake (not drinking enough water)
- Poor circulation
- Some drugs (oral contraceptives are a classic example)
- Pants that are too tight and high heels (they impede circulation and cause cellulite)

Discussion: The method that I propose below can be performed by anyone but is particularly designed for people with flabby thighs and cellulite. It aims to

- restore the circulation in areas of stagnation by draining fluids, helping to mobilize the adipose tissue,
- improve the alignment of the legs by balancing the myofascial lines,
- restore the body's balance (upper body, lower body),
- improve the pelvic floor, and
- improve posture.

Flabby thighs are often associated with an unbalanced Lateral Line and Deep Front Line, as well as a weak pelvic floor.

Focus On: What should a training session consist of? Training to combat flabby thighs should find the right balance between cardiovascular exercises, myofascial exercises, and muscle-strengthening exercises. For this there is nothing better than circuit training that includes

- aerobic exercises to promote lipolysis,
- myofascial exercises to restore the body's balance, and
- muscle exercises to tone the muscles.

This approach tackles flabby thighs and cellulite on all fronts.

Concept: Non-stop program and sequence of exercises.

Structure: Circuit training, a sequence of 6 to 15 exercises (training program) performed non-stop or with short breaks.

Focus On: Balancing tension in the leg lines LL–DFL–SBL–SFL. Improves: Posture, functionality of the microcirculation, and body shape.

Strategies: RELEASE, FEEL IT, MOBILITY, STRETCH, and ENERGY.

CIRCUIT TRAINING – BASIC PROGRAM
EQUIPMENT: SENSORY BALL – MINI BAND – FOAM ROLLER **Goal:** To combat flabby thighs

Warm-up
Cardio 5 min
FOOT MASSAGE WITH BALL
1 x 20 sec per foot (page 60)

Concentric ON/OFF
SHORTENED HIP FLEXOR
1 x 3 x 3 sec per leg (page 90)

Eccentric ON/OFF HIP FLEXOR STRETCH
1 x 3 x 3 sec per leg (page 91)

STEPPING BACK AND FORTH
1 x 10 per side (page 109)

BEAT IN STANDING POSITION
1 x 20 sec on the inner thigh and problem areas (page 92)
Cardio 5 min

Middle phase
KNEE SAVER-LATERAL STABILIZATION
1 x 10 per side (page 120)

THREE-LEGGED WINDMILL AGAINST THE WALL 1 x 5 per side (page 210)
Front squat 1 x 5
Push-up 1 x 5–15

ELASTIC BOUNCES 3 x 10 sec (page 237)
Cardio 1 min

Repeat non-stop 2 or 3 times

Active recovery 3 to 4 min:
ZIG ZAG ON THE LATERAL LINE AGAINST THE WALL 1 x 3 per side (page 277)

SIDEWAYS MOVEMENTS WITH MINI BAND 2 x 5 per side (page 94)

FRONTAL WOODPECKER: MOVEMENTS IN HALF-KNEELING POSITION
1 x 5 per side (page 114)

BRIDGE AGAINST THE WALL
1 x 5 per leg (page 140)

Suspension training Low Row
1 x 10–15
ABDOMINAL X 1 x 10 per side (page 58)
Cardio 1 min

Repeat non-stop 2 or 3 times

Active recovery 3 to 4 min:
BEAT IN STANDING POSITION
1 x 30 sec on the whole thigh (page 92)

Cool-down
SIDEWAYS ROCKING ON THE SACRUM WITH A FOAM ROLLER
1 x 30 sec (page 269)

ZIG ZAG ON THE LATERAL LINE WITH A SENSORY BALL
1 x 3 per side (page 277)

BALANCED BREATHING
1 x 1 min (page 286)
Cardio 5 to 10 min

Training Guidelines
Perform the exercises barefoot.
Keep breathing (do not hold your breath).
Training: 2 or 3 times a week for 2 weeks and then move on to the next program.
Include a 30 to 60 minute walk every other day.
Homework: Beat in Standing Position and Elastic Bounces 3 or 4 times during the day.

Increase the Intensity
Replace the Bridge Against the Wall exercise with the gymball leg curl 1 x 5

Note: You can also include the exercises from the various programs in other types of training to fully meet your training needs.

CIRCUIT TRAINING – INTERMEDIATE PROGRAM
EQUIPMENT: FOAM ROLLER – CLUBS – WEIGHT **Goal**: To combat flabby thighs and cellulite

Warm-up
Cardio 5 min
ANTERIOR STIMULATION OF THE ANKLES RESTING ON THE FOAM ROLLER 1 x 10 (page 95)

PRESSURE AND MASSAGE ON THE CALF MUSCLE WITH A FOAM ROLLER 1 x 5 per leg (page 98)

PRESSURE AND MASSAGE ON THE INNER LEG WITH A FOAM ROLLER 1 x 5 per leg (page 99)

STEPPING BACK AND FORTH + UPPER BODY ROTATIONS 1 x 10 per side (page 110)

SIDE STEP AND DIAGONAL STEP 1 x 10 per side (page 111)

Middle phase
SIDEWAYS LUNGE 1 x 5 per leg (page 112)

SWING WITH WOODEN CLUBS 1 x 10 (page 255)

GECKO PUSH-UP AGAINST THE WALL 1 x 5–10 (page 250)

CRAB 1 x 5–10 (page 168)

DIAGONAL LUNGE 1 x 5 per leg (page 112)

SWING WITH WOODEN CLUBS 1 x 10 (page 255)

Repeat non-stop 2 to 4 times

Active recovery 3 to 4 min:
ELEPHANT IN MOTION 1 x 5 per leg (page 157)

BEAT IN STANDING POSITION 1 x 30 sec on the whole thigh (page 92)

BIRD DOG AGAINST THE WALL 1 x 30 sec per leg (page 176)

ROTATING SPLIT SQUAT AGAINST THE WALL 1 x 5 per leg (page 206)

BAMBOO 1 x 5 per side (page 180)

Push Press 1 x 5–10
Rowing with kettlebell 1 x 5–10
DYNAMIC LUNGE STARTER 1 x 5 per side (page 235)

Repeat non-stop 2 to 4 times

Active recovery 3 to 4 min:
FLASH ACTIVATION 1 x 20 sec (page 93)

SIDEWAYS ELASTIC BOUNCES 1 x 20 sec (page 238)

BEAT IN STANDING POSITION 1 x 30 sec on the whole thigh (page 92)

Cool-down
SUPINE CORKSCREW 1 x 5 per side (page 196)

PRONE CORKSCREW 1 x 3 per side (page 199)

ZIG ZAG ON THE LATERAL LINE WITH A SENSORY BALL 1 x 3 per side (page 277)

BALANCED BREATHING 1 x 1 min (page 286)
Cardio 5 to 10 min

Training Guidelines
Perform the exercises barefoot.

Keep breathing (do not hold your breath).

Training: Twice a week for 2 weeks and then move on to the next program.

Include a 30- to 60-minute walk every other day.

Homework: Beat in Standing Position and Elastic Bounces 3 or 4 times during the day.

Increase the Intensity
Sideways lunges with weights

Note: You can also include the exercises from the various programs in other types of training to fully meet your training needs.

CIRCUIT TRAINING – ADVANCED PROGRAM 1

EQUIPMENT: WEIGHTS – JUMP ROPE – SENSORY BALL
Goal: To combat flabby thighs and cellulite

Warm-up
Cardio 5 min
STORK AGAINST THE WALL
1 x 30 (page 82)

GLUTEAS MAXIMUS ON/OFF
AGAINST THE WALL
1 x 5 per side (page 83)

UNCOMFORTABLE HALF-
SITTING POSITION
1 x 5 per side (page 169)

ELEPHANT IN MOTION
1 x 5 per side (page 157)

THREE FEET IN MOTION
1 x 5 per leg (page 163)

SKIPPING 1 x 30 sec (page 239)

CIRCLES WITH YOUR ARM
AGAINST THE WALL
1 x 3 per arm (page 138)

CRAB 1 x 5–10 (page 168)

Middle phase
ELASTIC LUNGE 1 x 3–5 per side
(page 236)

SWING WITH DUMBBELLS
1 x 5 per side (page 257)

SKIPPING 1 x 30 sec (page 239)

High Row & W Position 1 x 10
ABDOMINAL X 1 x 10 per side
(page 58)

SKIPPING 1 x 30 sec (page 239)

Repeat non-stop 2 or 3 times

Active recovery 2-3 min:
FOOT MASSAGE WITH BALL
1 x 20 sec per foot (page 60)

SIDEWAYS DOG ON THREE
LEGS AGAINST THE WALL
1 x 5–10 per side (page 252)

Snatch with kettlebell
1 x 10 per arm
ROLLING BEETLE 1 x 5 per side
(page 218)

SKIPPING 1 x 30 sec (page 239)

SYNCHRONIZED SWINGING
WALKING LUNGE WITH
KETTLEBELLS
1 x 7 per leg (page 259)

Floor pull 1 x 5–10
KNEE SAVER WITH ROTATION
1 x 5–10 per side (page 121)

SKIPPING 1 x 30 sec (page 239)

Repeat non-stop 2 or 3 times

Active recovery 2-3 min:
BEAT IN STANDING POSITION
1 x 30 sec on the whole thigh
(page 92)

Cool-down
THREE-LEGGED SCORPION
STINGER (page 187)
Cardio 5 to 10 min

Training Guidelines
Perform the exercises barefoot.

Keep breathing (do not hold your breath).

Training: 2 or 3 times a week for 2 weeks and then move on to the next program.

Include a 30- to 60-minute walk every other day, increasing the distance.

Homework: Flash Activation and Foot Massage With Ball twice during the day.

Note: You can replace skipping with a cardio machine. You can also include the exercises from the various programs in other types of training to fully meet your training needs.

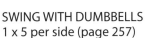
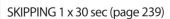

CIRCUIT TRAINING – ADVANCED PROGRAM 2

EQUIPMENT: RESISTANCE BAND – CLUBS – JUMP ROPE – SENSORY BALL
Goal: To combat flabby thighs and cellulite

Warm-up
Cardio 10 min

Middle phase
HOPSCOTCH 1 x 30 sec
(page 238)

BEAT IN STANDING POSITION
1 x 15 sec on the whole thigh
(page 92)

ABDOMINALS AGAINST THE
WALL 1 x 7 (page 179)

Assisted pull up with resistance
band 1 x 5–10
SKIPPING 1 x 30 sec (page 239)

Overhead squat 1 x 5–7
SIDEWAYS DOG ON THREE
LEGS AGAINST THE WALL
1 x 5–10 per side (page 252)

KNEELING BEAT 1 x 15 sec on
the whole thigh (page 92)

SIDEWAYS MONKEY
1 x 5 per side (page 242)

AIRCRAFT MARSHALLING WITH
CLUBS 1 x 10 (page 142)

SKIPPING 1 x 30 sec (page 239)

KNEE SAVER-LATERAL
STABILIZATION 1 x 10 per side
(page 120)

SUPPORTED LUNGE WITH BOW
1 x 5–10 per leg (page 235)

Repeat 1 to 3 times

BEAT IN STANDING POSITION
1 x 15 sec on the whole thigh
(page 92)

CRICKET PUSH-UP
1 x 5 (page 247)

CRAB 1 x 5–10 (page 168)

SKIPPING 1 x 30 sec (page 239)
Recovery: 3 to 4 min

Cool-down
Cardio 10 min
FOOT MASSAGE WITH BALL
1 x 20 sec per foot (page 60)

TWISTING SCORPION
1 x 3–5 per side (page 216)

Training Guidelines
Perform the exercises barefoot.

Training: 2 or 3 times a week for 2 weeks and then move on to the next program.

Include a 30- to 60-minute walk every other day, increasing the distance.

Homework: Breathe deeply whenever your phone rings or you receive a text.

Note: You can replace skipping with a cardio machine. You can also include the exercises from the various programs in other types of training to fully meet your training needs.

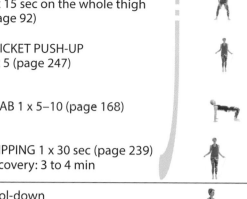

9.6. PROGRAM TO IMPROVE RUNNING PERFORMANCE

Concept: Non-stop program and sequence of exercises.

Structure: One sequence of 6 to 15 exercises constitutes a training program.

Focus On: Improving joint functionality and running performance.

Strategies: RELEASE, STRETCH, ENERGY, and MOBILITY.

RUNNING
EQUIPMENT: SENSORY BALL **Goal**: Warm-up – Cool-down

Warm-up
CIRCLES WITH YOUR ARM AGAINST THE WALL 1 x 10 per arm (page 138)

Circles with your wrists 1 x 10 per wrist
ELASTIC BOUNCES 3 x 10 sec
(page 237)

SHIFT SQUAT 1 x 5 per side
(page 181)

BEAT IN STANDING POSITION
1 x 30 sec on the whole leg (page 92)

STEPPING BACK AND FORTH
1 x 5 per leg (page 109)

"L" STEP 1 x 5 per leg (page 110)

ROUND ELEPHANT
1 x 10 per leg (page 156)

SWING FROM LOW TO HIGH
1 x 10 (page 240)

Middle phase
During your run, perform elastic bounces for 30 seconds every 1.25 miles (2 km) and then continue your run. You will instantly notice that after each active recovery, the explosiveness and speed of your running will increase as your fascial tissue is rehydrated, optimizing your performance.

Cool-down
Do the foam roller exercises on your entire leg to speed up your recovery.

FOOT MASSAGE WITH BALL
1 x 20 sec per foot (page 60)

Training Guidelines
Perform the exercises barefoot.
Perform the exercises with precision.
Note: You can also include the exercises from the various programs in other types of training to fully meet your training needs.

9.7. PROGRAMS TO IMPROVE MOBILITY AND FLEXIBILITY

Concept: Non-stop program and sequence of exercises.

Structure: One sequence of 6 to 15 exercises constitutes a training program.

Focus On: Improving functional mobility.

Strategies: MOBILITY and STRETCH.

Discussion: The programs that I propose below are designed to improve your joint mobility.

- Exercises for joint compartments
- Combination of mobility and stretch exercises

Integration in your training: Include exercises for specific joint systems in your training. Example: Before an intense arm workout, choose 2 to 4 shoulder, elbow, and wrist mobilization exercises.

MOBILITY – PROGRAM 1
EQUIPMENT: RESISTANCE BAND **Goal**: Mobility of the main joints

Functional mobility: ankle/foot
STEPPING BACK AND FORTH
WITH RAISED ARMS (page 109)

"L" STEP (page 110)

STEP WITH ROTATION
(page 110)

SAGITTAL WOODPECKER: MOVEMENTS IN HALF-KNEELING POSITION (page 113)

FRONTAL WOODPECKER: MOVEMENTS IN HALF-KNEELING POSITION (page 114)

CRAB IN SQUAT POSITION (page 117)

Functional mobility: hip
MOVING STORK AGAINST THE WALL (page 135)

KNEE SAVER-LATERAL STABILIZATION (page 120)

KNEE SAVER WITH STRAIGHT LEG (page 121)

CIRCLES WITH THE RESISTANCE BAND (page 141)

Functional mobility: spinal column
BRIDGE AGAINST THE WALL (page 140)

THORACIC SPINE ROTATIONS (page 130)

SIDEWAYS TILTING OF THE UPPER BODY IN STANDING POSITION (page 126)

Functional mobility: scapulohumeral joint
SPIDER WOMAN (page 131)

THE SHOULDER BLADES WRAP (page 133)

CIRCLES WITH YOUR ARM AGAINST THE WALL (page 138)

Training Guidelines

Perform the exercises barefoot.

Perform the exercises with precision.

Perform 5 to 8 repetitions per exercise.

Training: 3 times a week.

Note: You can also include the exercises from the various programs in other types of training to fully meet your training needs.

MOBILITY – PROGRAM 2

EQUIPMENT: CLUBS – KETTLEBELL
Goal: Mobility of the main joints

Functional mobility: ankle/foot

STEP WITH ROTATION (page 110)

SIDEWAYS LUNGE (page 112)

DIAGONAL LUNGE (page 112)

SPREADING YOUR KNEES IN HALF-SITTING POSITION (page 115)

GOOSE WALK IN HALF-SITTING POSITION (page 116)

Functional mobility: hip

KNEE SAVER WITH ROTATION (page 121)

BEACHED STORK - KNEE RAISE (page 123)

TRANSVERSE SEATED WIND-SHIELD WIPERS (page 125)

Functional mobility: spinal column
SIDEWAYS TILTING OF THE UPPER BODY IN STANDING POSITION (page 126)

TIRED CAT ON ALL FOURS (page 128)

THORACIC SPINE ROTATIONS ON TWO POINTS OF SUPPORT (page 129)

BRIDGE AGAINST THE WALL (page 140)

Functional mobility: scapulohumeral joint

SPIDER WOMAN (page 131)

CIRCLES WITH YOUR ARM AGAINST THE WALL (page 138)

AIRCRAFT MARSHALLING WITH CLUBS (page 142)

INFINITE SWING WITH KETTLEBELL (page 260)

Training Guidelines

Perform the exercises barefoot.

Perform the exercises with precision.

Perform 5 to 8 repetitions per exercise.

Training: 3 times a week.

Note: You can also include the exercises from the various programs in other types of training to fully meet your training needs.

MOBILITY AND STRETCH – PROGRAM 3

Goal: Mobility and flexibility

STEPPING BACK AND FORTH (page 109)

ELEPHANT (page 155)

UNCOMFORTABLE HALF-SITTING POSITION (page 169)

FRONTAL WOODPECKER: MOVEMENTS IN HALF-KNEELING POSITION (page 114)

SIDEWAYS LUNGE (page 112)

DIAGONAL LUNGE (page 112)

THREE-LEGGED SCORPION STINGER (page 187)

KNEE SAVER WITH ROTATION (page 121)

ROLLING BEETLE (page 218)

TRANSVERSE SEATED WIND-SHIELD WIPERS (page 125)

WINDMILL IN HALF-KNEELING POSITION (page 184)

CAMEL WALK (page 159)

SPIDER WOMAN (page 131)

SHIFT SPLIT SQUAT (page 182)

SPREADING YOUR TOES (page 166)

THREE FEET IN MOTION (page 161)

THE SHOULDER BLADES WRAP (page 133)

SPREADING YOUR KNEES IN HALF-SITTING POSITION (page 115)

GOOSE WALK IN HALF-SITTING POSITION (page 116)

BRIDGE (page 171)

COBRA (page 173)

SUPINE CORKSCREW IN ACTION (page 198)

ROTATING COBRA (page 200)

LIZARD (page 189)

BAMBOO WITH LEGS CROSSED (page 127)

Training Guidelines

Perform the exercises barefoot.
Perform the exercises fluidly and with precision.
Perform 3 to 6 repetitions per exercise or hold for 15 to 20 sec
Training: 2 or 3 times a week.

STRETCH

Goal: To improve flexibility

Concept: Non-stop program and sequence of exercises.
Structure: One sequence of 6 to 15 exercises constitutes a training program.
Focus On: Improving flexibility.
Strategies: STRETCH and FEEL IT.

BEAT IN STANDING POSITION
1 x 20 sec on the inner thigh and problem areas (page 92)

ELASTIC BOUNCES 3 x 20 sec
(page 237)

FLASH ACTIVATION 1 x 20 sec
(page 93)

ELEPHANT IN MOTION
1 x 5 per leg (page 157)

ROUND ELEPHANT
1 x 10 per leg (page 156)

SHIFT SQUAT 1 x 5 per side
(page 181)

UNCOMFORTABLE HALF-SITTING POSITION 3 x 5 per leg
(page 169)

ELEPHANT IN MOTION
1 x 5 per side (page 157)

THREE FEET IN MOTION
1 x 4 per side (page 162)

THREE FEET IN MOTION
5 per leg (page 162)

OPENING A BOOK 1 x 4 per side
(page 203)

ROLLING BEETLE 1 x 4 per side
(page 218)

BEETLE STRETCH 1 x 4
(page 218)

BRIDGE CORKSCREW
1 x 4 per side (page 195)

THREE-LEGGED SCORPION STINGER 3 per side (page 187)

WINDMILL IN HALF-KNEELING POSITION 1 x 4 per side
(page 184)

THREE-LEGGED WINDMILL
1 x 4 per side (page 186)

SUPINE CORKSCREW
1 x 4 per side (page 196)

PRONE CORKSCREW
1 x 4 per side (page 199)

ATTACKING SCORPION
1 x 3 per side (page 167)

STANDING CORKSCREW
1 x 3 per side (page 183)

LIZARD 1 x 20 sec per side
(page 189)

DYNAMIC CAMEL 1 x 5
(page 160)

Return to a standing position

Training Guidelines

Perform the exercises barefoot.

Perform the exercises correctly and in a controlled manner.

Training: 3 times a week.

Note: You can also include the exercises from the various programs in other types of training to fully meet your training needs.

9.8. PROGRAMS FOR PILATES

PILATES – BASIC PROGRAM

EQUIPMENT: SENSORY BALL – RESISTANCE BAND
Goal: To improve the execution of the exercises

Concept: Non-stop program and sequence of exercises.
Structure: One whole sequence of exercises constitutes a training program.
Focus On: improving flexibility, mobility, stability, and toning.
Strategies: FEEL IT, MOBILITY, STRETCH, RELEASE, and ENERGY.

ELASTIC BOUNCES 3 x 10 sec
(page 237)

BACK TO THE WALL IN STANDING POSITION 3 x 20 sec
(page 81)

BACK TO THE WALL WITH ARMS RAISED 1 x 10 (page 81)

KNEE FALL OUT WITH A SENSORY BALL (page 268)

The Hundred
The Roll up
CIRCLES WITH THE RESISTANCE BAND 1 x 10 per side
(page 141)

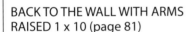

The One Leg Circle
Rolling Back
The One Leg Stretch
The Spine Stretch
CHEST LIFT IN PRONE POSITION WITH A SENSORY BALL 1 x 3–5
(page 273)

LEG LIFT IN PRONE POSITION WITH A SENSORY BALL
3 per position (page 272)
The Swan-Dive

Training Guidelines
Perform the exercises barefoot.

Perform the exercises correctly and in a controlled manner.

Training: 3 times a week.

PILATES – INTERMEDIATE PROGRAM

EQUIPMENT: SENSORY BALL
Goal: To improve the execution of the exercises

FOOT MASSAGE WITH BALL
1 x 30 sec per foot (page 60)

EYES ALERT, NECK LOOSE
(page 87)

The Hundred
The Roll up
SITTING ON A SENSORY BALL
IN THE "Z" POSITION (page 276)

The One Leg Circle
Rolling Back
The One Leg Stretch
The Spine Stretch
THREE FEET IN MOTION
1 x 5 per side (page 161)

Rocker With Open Legs
The Saw
The One Leg Kick
BOOMERANG IN SIDEWAYS
HALF-KNEELING POSITION
1 x 5 per side (page 188)

The Side Kicks
SCREWDRIVER (page 202)

The Spine Twist
DYNAMIC CAMEL 1 x 5 with platform under your hands
(page 160)

Push-up
SWING FROM LOW TO HIGH
1 x 10 (page 240)

9.9. PROGRAMS FOR FUNCTIONAL TRAINING

FUNCTIONAL TRAINING
EQUIPMENT: WEIGHTS **Goal**: Full-body workout

Warm-up
Cardio 5 min
BEAT IN STANDING POSITION
1 x 30 sec (page 92)

SIDE LUNGE WITH MOVEMENT
OF THE EYES (page 88)

THE SHOULDER BLADES WRAP
(page 133)

KNEE SAVER-LATERAL
STABILIZATION 1 x 5 (page 120)

KNEE SAVER 1 x 5 (page 121)

THREE FEET IN MOTION 5 x leg
(page 161)

SIDEWAYS SPIDER AGAINST THE
WALL 1 x 5 sec (page 222)

SHIFT SPLIT SQUAT 1 x 5 per leg
(page 182)

CIRCLES WITH YOUR ARM
AGAINST THE WALL
1 x 10 per arm (page 138)

DIAGONAL LUNGE 1 x 5 per leg
(page 112)

SWING WITH DUMBBELLS
1 x 5–7 per side (page 257)

DOG BREATHING (PANTING)
1 x 30–45 sec (page 86)

Middle phase
Squat 1 x 10
ELASTIC BOUNCES
1 x 30 sec (page 237)
Repeat 3 times non-stop

Push-up 1 x 10
SPIDER WOMAN 1 x 5
(page 131)
Repeat 3 times non-stop

ABDOMINAL X 1 x 10 per side
(page 58)

THREE FEET IN MOVEMENT
1 x 10 per side (page 163)
Repeat 3 times non-stop

Walking lunges with KB
1 x 5 per leg
BEAT 1 x 30 sec (page 92)
Repeat 3 times non-stop

Rowing KB 1 x 10
GIRAFFE (page 225)
Repeat 3 times non-stop

ABDOMINALS AGAINST THE
WALL 1 x 10 (page 179)

BIRD DOG AGAINST THE WALL
1 x 30–45 sec (page 176)
Repeat 3 times non-stop

Cool-down
ELASTIC BOUNCES 1 x 30 sec
(page 237)

SWING FROM LOW TO HIGH
1 x 10 (page 240)

BALANCED BREATHING
1 x 1 min (page 286)

Training Guidelines
Pairs of exercises in which each exercise follows a fascia workout.

Note: You can also reverse the sequence of paired exercises, performing the fascia exercise first followed by the isotonic (or anisometric) exercise.

FUNCTIONAL TRAINING PREPARATORY EXERCISES

EQUIPMENT: MEDICINE BALL - RESISTANCE BAND
Goal: To improve the execution of the exercise

Squat preparation:

KNEE SAVER-LATERAL STABILIZATION 1 x 10 per side (page 120)

STORK AGAINST THE WALL 5 x 3 sec per leg (page 82)

SHIFT SQUAT 1 x 5 per side (page 181)

ROLLING BEETLE (page 218)

SHIFT SQUAT 1 x 5 per side (page 181)

Lunge preparation:

ON/OFF SHORTENED HIP FLEXOR ON YOUR BACK (page 90)

Eccentric ON/OFF HIP FLEXOR STRETCH (page 91)

GLUTEUS MAXIMUS ON/OFF WITH MEDICINE BALL (page 84)

One Leg Deadlift preparation:

ONE LEG DEADLIFT AGAINST THE WALL (page 136)

ONE LEG DEADLIFT WITH RESISTANCE BAND (page 137)

THREE FEET IN MOTION AGAINST THE WALL (page 165)

Windmill preparation:

WINDMILL AGAINST THE WALL (page 208)

DYNAMIC LITTLE MERMAID (page 193)

Press preparation:

ARMS TO THE WALL (page 79)

SPIDER AGAINST THE WALL 3 x 15 sec (page 221)

BACK TO THE WALL WITH ARMS RAISED (page 81)

BIRD DOG AGAINST THE WALL (page 176)

GIRAFFE: SWINGING WITH RESISTANCE BAND (page 225) ALL SHOULDER, ARM, AND BACK EXERCISES WITH A BALL OR FOAM ROLLER

Plank preparation:

SPIDER WOMAN (page 131)

SIDEWAYS SPIDER AGAINST THE WALL (page 222) ALL SHOULDER, ARM, AND BACK EXERCISES WITH A BALL OR FOAM ROLLER

9.10. S.O.S. EXERCISES

Cramping during training? Some remedies:

BACK MASSAGE AGAINST THE WALL WITH A SENSORY BALL (page 270)

BACK MASSAGE ON THE FLOOR WITH A SENSORY BALL (page 270)

Training Guidelines
Loosen up: Stay still on a painful trigger point, continue to breathe, and apply gentle pressure on the ball every time you breathe out until the tension subsides, then move half an inch (1 cm) in the direction of your heart and repeat. On certain trigger points you could also move in small circles or draw a star.

9.11. POST-TRAINING EXERCISES

Example 1:

Recovery (after intense training of the arms)
FOREARM MOVEMENTS ON ALL FOURS WITH A FOAM ROLLER (page 282)

ARM MOVEMENTS ON ALL FOURS WITH A FOAM ROLLER (page 283)

Example 2:

Recovery (after intense training of the legs)
LEG LIFT IN PRONE POSITION WITH BALL 1 x series per side (page 272)

ZIG ZAG ON THE LATERAL LINE WITH A SENSORY BALL (page 277)

ROCKING ON THE SACRUM WITH A SENSORY BALL 1 x 5–10 (page 267)

CIRCLES AROUND THE PIRIFORMIS WITH THE DUO BALL (page 275)

ANTERIOR STIMULATION OF THE ANKLES WITH A FOAM ROLLER (page 95)

PRESSURE AND MASSAGE ON THE CALF MUSCLE WITH A FOAM ROLLER (page 98)

PRESSURE AND MASSAGE ON THE INNER LEG WITH A FOAM ROLLER (page 99)

BEAT (HAVE FUN TAPPING) (page 92)

9.12. FReE PROGRAMS

FReE – PROGRAM 1
Goal: To rejuvenate the fascial structures

Concept: Non-stop program and sequence of exercises.
Structure: One whole sequence of exercises constitutes a training program.
Focus On: Hydrating the fascial tissue and keeping it young.
Strategies: FEEL IT, MOBILITY, STRETCH, RELEASE, and ENERGY.

FOOT MASSAGE 45 sec
(page 60)

STEP WITH ROTATION 1 x 10
(page 110)

ELASTIC BOUNCES 3 x 30 sec
(page 237)

STANDING SWING FROM LOW TO HIGH 1 x 10 (page 240)

BAMBOO WITH LEGS CROSSED 1 x 5 (page 127)

EYES ALERT AND REACTIVE WITH SIDEWAYS LUNGE 1 x 5 (page 88)

DIAGONAL LUNGE 1 x 5 per leg (page 112)

CRAB IN SQUAT POSITION 1 x 3-5 per side (page 117)

RUNNER'S KNEE IN SQUATTING POSITION 1 x 5 (page 118)

SPIDER WOMAN 1 x 5 (page 131)

SPREADING YOUR KNEES IN HALF-SITTING POSITION 1 x 5 (page 115)

KNEE SAVER WITH ROTATION 1 x 10 per side (page 121)

TRANSVERSE SEATED
WINDSHIELD WIPERS
1 x 5 per side (page 125)

CAMEL WALK 1 x 10 (page 159)

THE SHOULDER BLADES WRAP
1 x 5 per side (page 133)

ATTACKING SCORPION
1 x 5 per side (page 167)

END TO END SIDEWAYS
SWINGS WITH CHANGE
OF TRAJECTORY
1 X 5-10 per side (page 241)

CRAB 1 x 5 (page 168)

THORACIC SPINE ROTATIONS
ON TWO POINTS OF SUPPORT
1 x 5 per side (page 129)

ROUND ELEPHANT
1 x 5 per side (page 156)

ELEPHANT IN MOTION
1 x 5 per side (page 157)

SUPPORTED LUNGE WITH BOW
1 x 5 per side (page 235)

BOOMERANG IN SIDEWAYS
HALF-KNEELING POSITION
1 x 4 per side (page 188)

DYNAMIC CAMEL 1 x 5
(page 160)

FLYING ATTACKING SCORPION
1 x 4 per leg (page 244)

SHIFT SQUAT IN HALF-
KNEELING POSITION
1 x 4 per side (page 182)

STANDING CORKSCREW
1 x 4 per side (page 183)

THREE-LEGGED WINDMILL
1 x 4 per side (page 186)

GORILLA 1 x 5 (page 214)

JUMPING DYNAMIC FROG 1 x 4
(page 215)

SHIFT SPLIT SQUAT
1 x 5 per side (page 182)

END TO END SIDEWAYS
SWINGS WITH CHANGE OF
TRAJECTORY
1 x 5 per side (page 241)

SIDEWAYS MONKEY
1 x 3 per side (page 242)

SITTING UNCOMFORTABLY ON
BOTH YOUR HEELS 1 x 3
(page 169)

LIZARD 1 x 3 per side
(page 189)

DYNAMIC LITTLE MERMAID
1 x 3 per side (page 193)

SUPINE CORKSCREW
1 x 3 (page 196)

SUPINE CORKSCREW IN ACTION
1 x 3 per side (page 198)

PRONE CORKSCREW
1 x 3 per side (page 199)

5–10 BALANCED BREATHING
WITH WEIGHT (page 285)

Training Guidelines
Perform the exercises barefoot.

Perform the exercises correctly, and in a controlled
and elastic manner.

Training: 2 or 3 times a week.

FASCIA-NATING MOVES FReE BEAST
Goal: To move freely

Concept: Non-stop program and sequence of exercises.
Structure: One whole sequence of exercises constitutes a training program.
Focus On: keeping the tissue hydrated and elastic.
Strategies: MOBILITY, STRETCH, and ENERGY.

Training Guidelines
Perform the exercises barefoot.
Perform 3 to 6 repetitions per exercise and side.
Perform the exercises fluidly and elastically.
Training: 2 or 3 times a week.

EXERCISE INDEX

BIBLIOGRAPHY

[1]   Albini E. (2010) *Pilates Woodpole*. Red, Milano

[2]   Albini, E. (2008) *Pilates per tutti*. Elika, Cesena

[3]   Albini, E. (2010) *Pilates per lo sport*. Red, Milano

[4]   Bruscia, G. (2008) *Donne in palestra: l'allenamento giusto*. Elika, Cesena

[5]   Bruscia, G. (2010) *Addominali per tutti*. Elika, Cesena

[6]   Bruscia, G. (2010) *Kettlebell*. Elika, Cesena

[7]   Bruscia, G. (2017) *La verità sull'allenamento funzionale*. Elika, Cesena

[8]   Buckminster Fuller, R. (1975) *Synergetics*. New York: Macmillan

[9]   Calais-Germain, B. (2015) *Anatomia per il movimento. Introduzione all'analisi delle tecniche corporee*. Epsylon, Roma

[10]   Canepa, L., "Il benessere visivo. Il metodo Bates" [http://www.benessere.com/salute/arg00/metodo_bates.htm, consulted in March 2018]

[11]   Chetta, G (2010) "Propriocettori" [http://www.giovannichetta.it/propriocettori.html, consulted in March 2018]

[12]   Cook, G. (2010) *Functional Movement Systems*. On Target Publications, California

[13]   Denys-Stuyf, G. (1996) *Il manuale del mézierèrista*. Marrapese, Roma

[14]   Dipartimento di Scienze della vita e Biologia dei sistemi. "Connettivi propriamente detti", in Atlante di Citologia ed Istologia [http://www.atlanteistologia.unito.it/page.asp?xml=connettivi.connettivi%20propriamente%20detti, consulted in March 2018]

[15]   Douglas, J. (1707) *Myographiae Comparatae Specimen*. Printed by W.B. for G. Strachan, London

[16]   Earls, J. (2014) *Born to Walk*. Berkeley: North Atlantic

[17]   Fascia Research Congress 2009, Boston [http://www.fasciacongress.org/2009]

[18]   Frederick, C., Frederick, A. (2016) *Faszienstretching: Diagnose, Behandlung, Training*. Riva. Kindle edition, 58

[19]   Freiwald, J. (2009) *Optimales Dehnen. Sport-Prävention-Rehabilitation*. Balingen: Spitta Verlag

[20]   Freiwald, J. (2013) *Optimales Dehnen. Sport-Prävention-Rehabilitation* (2nd Ed.). Balingen: Spitta

[21]   Freiwald, J. (2014) *Effectiveness of Adjuvant (Supplementary) ThermaCare® Heat Packages in the Treatment of Chronic Low Back Pain Patients in a Multimodal Setting*. Wuppertal: Bergische Universität Wuppertal

[22]   Freiwald, J., Baumgart, C., Konrad, P. (2007) *Einführung in die Elektromyographie. Sport-Prävention-Rehabilitation*. Balingen: Spitta

[23] Freiwald, J., Greiwing, A. (2014) *Optimales Krafttraining. Sport-Rehabilitation-Prävention.* Balingen: Spitta

[24] Frigo, G. (2008) *Anatomia e funzionalità della fascia crurale*, Thesis Università di Padova [http://tesi.cab.unipd.it/27901/1/ANATOMIA_E_FUNZIONALITA.pdf, consulted in February 2018]

[25] Garfin, S.R., Tipton, C.M., Mubarak, S.J., Woo, S.L., Hargens, A.R, Akeson, W.H. (1981) "Role of fascia in maintenance of muscle tension and pressure". J Appl Physiol, 51: 317-320

[26] Hegge, J. (1993) "Alcune riflessioni sul rapporto tra visione, proprlocezione e cinetica". J of Behav Optometry, 4: 95-97

[27] Help Chemistry, Sacra chimica blog (2017) "Osmosi e osmosi inversa" [https://sacrachimicablog.wordpress.com/2017/03/27/osmosi-e-osmosi-inversa-2/, consulted in March 2018]

[28] Hinz, B., Gabbiani, G., Chaponnier, C. (2002) "The NH2-terminal peptide of α–smooth muscle actin inhibits force generation by the myofibroblast in vitro and in vivo". Journal of Cell Biology, 13; 157(4): 657–663; doi: 10.1083/jcb.200201049

[29] Hinz, B., Phan, S.H., Thannickal, V.J., et al. (2012) "Recent developments in myofibroblast biology: paradigms for connective tissue remodeling". Am J Pathol, 180: 1340-1355

[30] Ingber, D.E. et al. (1993) "Mechanotransduction across the cell surface and through the cyto-skeleton". Science, 260: S. 1124-1127

[31] Jarvinen, T.A., Jozsa, L., Kannus, P., Jarvinen, T.L., Jarvinen, M. (2002) "Organization and distribution of intramuscular connective tissue in normal and immobilized skeletal muscle. An immunohistochemical, polarization and scanning electron microscopic study". J Musc Res Cell Mot, 23: 245-254

[32] Kapandji, I.A. (2004) *Fisiologia articolare.* Monduzzi Editore, Bologna

[33] Kawakami, Y., Muraoka, T., Ito, S., Kanehisa, H., Fukanaga, T. (2002) "In vivo muscle fibre behaviour during countermovement exercise in humans reveals a significant role for tendon elasticity". J Physio, 540: 635-646

[34] Laborit, H. (2000) *L'inhibition de l'action. Biologie, physiologie, psychologie, sociologie.* Paris-Montréal

[35] Langevin, H.M. (2006) "Connective tissue: a body-wide signaling network?" Med Hypotheses, 66(6):1074-7

[36] Langevin, H.M., Cornbrooks, C.J., Taatjes, D.J. (2004) "Fibroblasts form a body-wide cellular network". Epub, 122(1): 7-15

[37] Langevin, H.M., Fox, J.R., Koptiuch, C., et al. (2011) "Reduced thoracolumbar fascia shear strain in human chronic low back pain". BMC Musculoskelet. Disord, 19: 203

[38] Larsen, C. (2013) *Gut zu Fuss ein Leben lang.* Trias

[39] Leinonen, V., Kankaanpaa, M., Luukkonen, M., et al. (2003). "Lumbar paraspinal muscle function, perception of lumbar position, and postural control in disc herniation-related back pain". Spine, 28: 842-848

[40] Levine, P.A. (2002) *Traumi e shock motivi, come uscire dall'incubo.* Macro Edizioni

[41] Magnusson, P., Narici, M., Maganaris, C., Kjaer, M. (2008) "Human tendon behaviour and adaptation, in vivo". J Physiol, Jan 1; 586(Pt 1): 71-81; doi: 10.1113/jphysiol.2007.139105

[42] Masunaga, S. (2000) *Meridian Dehnuebungen*. Felicitas Huebner, Deutschland

[43] McGill, S. (2004) *Ultimate Back Fitness and Performance*. Waterloo, Ontario, Canada: Wabuno Publishers

[44] Medicina per tutti (2010). "Tessuto connettivo propriamente detto" [https://www.medicinapertutti.it/argomento/tessuto-connettivo-propriamente-detto/, consulted in March 2018]

[45] Medicina per tutti (2017). "Tessuto connettivo lasso" [https://www.medicinapertutti.it/argomento/tessuto-connettivo-lasso/, consulted in March 2018]

[46] Mobile Sport. "Siamo vecchi come le nostre articolazioni" [https://www.mobilesport.ch/actualita/riscaldamento-siamo-vecchicome-le-nostre-articolazioni/, consulted in March 2018 (source: "mobile" journal 4/2008, 20-21, Jost Hegner)]

[47] Mobile Sport. "Erwachsenensport Schweiz ESA, Mobile praxis, Magglingen" [https://www.mobilesport.ch/suchergebnisse/?q=fuss, consulted in March 2018]

[48] Myers, T.W. (2006) *Meridiani Miofasciali. Percorsi anatomici per i terapisti del corpo e del movimento.* Tecniche Nuove, Milan

[49] Myers, T.W. (2007) *Meridiani miofasciali.* Tecniche nuove, Milan

[50] Myers, T.W. (2014) *Anatomy Trains: Myofascial Meridians for Manual and Movement Therapists.* Health Sciences UK. Kindle edition

[51] Myers, T.W. (2014) *Anatomy Trains: Myofascial Meridians for Manual and Movement Therapists* (3rd Ed.). Edinburgh: Churchill Livingstone Elsevier

[52] Myers, T.W., Fredrick, C. (2012) "Stretching and fascia". In: Schleip, R., Findley, T., Chaitow, L., et al. editors, *Fascia, the Tensional Network of the Human Body,* Edinburgh: Churchill Livingstone Elsevier

[53] Pilates, J.H., Miller, W.J., Robbins, J. (2000) *The Millennium Edition: Return to Life Through Contrology and Your Health.* Presentation Dynamics, ISBN 1-928564-00-3

[54] Pollack, G.H. (2013) *The Fourth Phase of Water. Beyond Solid, Liquid and Vapor.* Ebner and Sons Publishers, Seattle, Washington

[55] Rotolo. J., Roll, R. (1988) *Dall'occhio al piede: una catena propriocettiva coinvolta nel controllo posturale. (Postura e andatura: sviluppo, adattamento e modulazione).* Elsevier Science Publishers, 155-164

[56] Schleip, R. (2011) "Principles of Fascia Fitness" [www.terrarosa.com.au, Issue 7]

[57] Schleip, R., Baker A. (2015) *Fascia in Sport and Movement,* Handspring Publishing Limited, Scotland

[58] Schleip, R., Duerselen, L., Vleeming, A., Naylor, I.L., Lehmann-Horn, F., Zorn, A., Jaeger, H., Klingler, W. (2012) "Strain hardening of fascia: static stretching of dense fibrous connective tissues can induce a temporary stiffness increase accompanied by enhanced matrix hydration". Journal of Bodywork and Movement Therapies, 16: 94e100

[59] Schleip, R., Findley, T.W., Chaitow, L., Huijing, P.A. (2012) *Fascia: the Tensional Network of the Human Body,* Churchill Livingstone Elsevier

[60] Schleip, R., Müller, D.G. (2012) "Training principles for fascial connective tissues: Scientific foundation and suggested practical

applications". Journal of Bodywork & Movement Therapies, 1-13

[61] Schleip, R. (2002) "Fascial plasticity – a new neurobiological explanation: part 1" and "Fascial plasticity – a new neurobiological explanation: part 2". Journal of Bodywork and Movement Therapies, 7(1): 11-19; 7(2): 104-116

[62] Slomka, G., Regelin, P. (2005) *Stretching – aber richtig!* Blv-Verlag

[63] Staubesand, J., Li, Y. (1996) "Zum Feinbau der Fascia cruris mit besonderer Berucksichtigung epi- und intrafaszialer Nerven". Manuelle Medizin, 34: 196-200

[64] Stecco, C. (2014) *Functional Atlas of the Human Fascial System*. Elsevier Health Sciences UK. Kindle edition

[65] Stecco, C. (2015) *Functional Atlas of the Human Fascial System*. Churchill Livingstone Elsevier

[66] Stecco, C., Porzionato, A., Macchi, V., Tiengo, C., Parienti, A., Aldegheri, R., Delmas, V., De Caro, R. (2006) "Histological characteristics of the deep fascia of the upper limb". Italian journal of anatomy and embryology, 111(2): 105-10

[67] Stecco, C., Porzionato, A., Stecco, A., Macchi, V., Day, J.A., De Caro, R. (2008) "Histological study of the deep fasciae of the limbs". Journal of Bodywork and Movement Therapies, 12(3): 225-230

[68] Stecco, L., Stecco C. (2008) *Fascial Manipulation*. Padua: Piccini Publisher

[69] Still, A.T. (1899) *Philosophy of Osteopathy*. Academy of Osteopathy, Kirksville, MO

[70] Tomasek, J.J., Gabbiani, G., Hinz, B., Chaponnier, C., Brown, R.A. (2002) "Myofibroblasts and mechano-regulation of connective tissue remodeling". Nat Rev Mol Cell Biol, 3(5): 349-363

[71] Treccani. "Hooke, Robert" [http://www.treccani.it/enciclopedia/robert-hooke/, consulted in March 2018]

[72] Vleeming, A., Mooney, V., Stoeckart, R. (2007) *Movement, Stability & Lumbopelvic Pain* (2nd Ed.). Churchill Livingstone Elsevier

[73] Wikipedia, "Sistema nervoso parasimpatico" [https://it.wikipedia.org/wiki/Sistema_nervoso_parasimpatico, consulted in March 2018]

[74] Wildman F. (2006) *The Busy Person's Guide to Easier Movement*. The Intelligent Body Press

[75] Wilke, J., Schleip, R., Klingler W., Stecco, C. (2017) "The Lumbodorsal Fascia as a Potential Source of Low Back Pain: A Narrative Review". Biomed Res Int, 2017: 5349620; doi: 10.1155/2017/5349620

[76] Willard, F. (2007) Fascial Continuity: Four Fascial Layers of the Body. Fascia Research Congress, Boston

[77] Willard, F.H., Vleeming, A., Schuenke., M.D., Danneels, L., Schleip, R. (2012) "The thoracolumbar fascia: anatomy, function and clinical considerations". J Anat, 2012 Dec; 221(6): 507-536; doi: 10.1111/j.1469-7580.2012.01511.x

[78] Wolf, C. (2003) "Flexibility Highways" [http://www.ptonthenet.com/articles/Flexibility-Highways-2032, consulted in March 2018]

[79] Wood, T.O., Cooke, P.H., Goodship, A.E. (1988) "The effect of exercise and anabolic steroids on the mechanical properties and crimp morphology of the rat tendon". Am J SportsMed, 16(2): 153-158

ABOUT THE AUTHOR

 **Ester Albini** is the owner of Ester Albini Pilates Academy and cofounder of the Functional Training School (FTS). She is a personal trainer and fitness instructor and has a diploma in fitness from the Swiss Confederation. Her professional training includes an impressive list of credentials, such as Anatomy Trains Levels I and II, IKFF Certified Kettlebell Trainer Level I, TRX Level II, Reebok Master Trainer, Polestar, Balanced Body, Gyrotonic Teacher Level 1, and Posturologa Mézières and Bricot. She has been a lecturer for several international schools, including FTS, the Italian Fitness Federation (FIF), Reebok, and the Swiss Academy of Fitness and Sports (SAFS). Albini is the creator of the Woodpole method, Body Ball Relaxing, and Fascial Real Emotion (FReE). She is also the author of several books and DVDs.

FReE & PILATES ACADEMY

FReE
Fascial REal Emotion

ITALY'S INAUGURAL MYOFASCIAL TRAINING SCHOOL

In almost 30 years of work, while I attained numerous qualifications abroad and worked as a teacher at various training institutions in Italy and Switzerland, I always thought that I would love to found my own school one day. "What would I like it to be like?" I thought. "Unique" was the answer. Completely innovative training where educational theory, preparation, and practice merge into a single path for the research and teaching of the perfect movement.

How can we define the perfect movement? Simply put, it is the key to opening all the other doors to well-being linked to physical activity: in a word, **intelligent** movement. Simple (but not banal!) exercises represent the starting point for those who want to learn how to move well and to look after themselves and their own body and who feel the need to share their path with others.

As I will explain, by working on the body more deeply than we are used to, we can understand things that are normally ignored and go unnoticed. By feeling my body and experimenting with a new way of using it, I discovered the meaning behind what I was doing and realized I had found a new key to movement.

My questions found an answer: **myofascial** movement.

I first encountered Guido Bruscia's Functional Training School (FTS) 10 years ago. It was a surprise, a training school that taught functional movement, the other side of the myofascial movement coin. I immediately felt a unity of identity and perspectives with the FTS.

My dream finally became a reality in 2012 when I founded the FReE & Pilates Academy, Italy's inaugural myofascial training school.

What Is It?

A training school unlike any other that offers training courses for professionals and enthusiasts, supporting them from the start until their qualification as professionals.

What Are Its Goals?

1. Provide unique and exclusive training courses
2. Create a constant link between school and student, offering students support and help throughout their journey

FReE & Pilates Academy Training Courses

- FReE (Fascial Real Emotion) 1st, 2nd, and 3rd level
- Myofascial Pilates:
    - o Matwork 1st, 2nd, and 3rd level
    - o Reformer 1st and 2nd level
    - o Studio (Cadillac and Chair) 1st and 2nd level
- Hamazon (myofascial training meets functional training in the training room) 1st and 2nd level

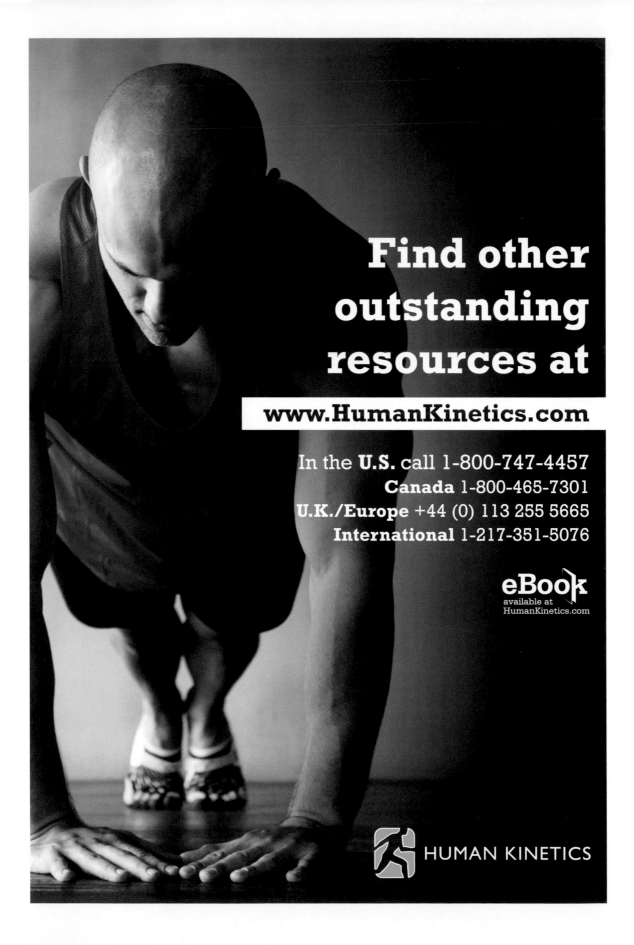